I.T. IN CRISIS
A New Business Model

L. Paul Ouellette

authorHOUSE®

AuthorHouse™
1663 Liberty Drive
Bloomington, IN 47403
www.authorhouse.com
Phone: 1-800-839-8640

First published by AuthorHouse 8/18/2009

ISBN: 978-1-4389-8544-2 (e)
ISBN: 978-1-4389-8542-8 (sc)
ISBN: 978-1-4389-8543-5 (hc)

Library of Congress Control Number: 2009906856

Printed in the United States of America
Bloomington, Indiana

This book is printed on acid-free paper.

Some people are never given a second chance for happiness. Marin, Ethan, and Linda have filled my life with many blessings and unconditional love. Linda was especially involved in every aspect of this book. I am very grateful.

WHY THIS BOOK IS IMPORTANT

IT, AND THE INFORMATION TECHNOLOGIST, are facing a crisis ... and "business as usual" can no longer sustain us.

For decades, we have been content to be the supplier of first choice when it comes to supplying computer technology solutions to our organizations. That is no longer enough. Today, IT as an organization needs to go far beyond being the technology experts.

Here's why: massive shifts in the economy have produced brand-new business models, and our organizations (both in the private and the public sectors) have had to adopt those models in order to stay competitive—or even to survive. If they are to endure, our organizations must create stronger, more reciprocal customer relationships then ever before—and so must IT.

Today, IT must become part of the *total business solution process.* Unfortunately, that is not where we are right now.

Let me share some of what high-level IT executives are saying about the challenges and transitions our teams must address today, tomorrow, and in the years ahead if we are to survive and be part of the solution.

The Gartner Group, a prestigious IT consulting firm, had this to say about the future of IT at a major technology conference held in October 2008:

- In recent years, enterprises have challenged their IT departments to increase their external focus *on customers,* as well as on new products and services, new geographies, and new business processes.
- Unfortunately, few CIOs have the staff with the *skills sets* to adequately meet these externally focused demands, and there has been little remaining funding for additional hires.
- Gartner recommends that CIOs recognize the skill gap, refrain from solely hiring staff with IT backgrounds in the

future, and focus on identifying and delivering distinctive, customized solutions for business—before it's too late.

To whom should these **business process professionals** report? Gartner recommends that they be oriented within a new, hybrid team, such as a business process competency center, that reports to a chief operating officer. In this scenario, competency centers would be made up of relatively few employees, but would be joined by the business domain's experts, process experts, and IT professionals for the duration of a project … only to return to their respective departments upon completion of the project.

That's not all. According to *Fortune* magazine …

- Technology spending as a percentage of Gross Domestic Product is expected to fall—not rise—for the first time in three decades that had seen continuous growth in this part of the economy.
- Venture capitalists invested $14.1 billion in 1,930 IT companies in 1998; fifty of them went public. They invested roughly the same amount in roughly the same number of companies in 2008; *two* went public. The fifty companies from 1998 posted revenues of $2.7 billion. The two IT companies who went public in 2008 only posted revenues of $249 million.
- More and more business experts are citing a move away from investments in familiar IT firms—and toward firms in other, more promising fields (such as clean energy and technology). (Source: *Fortune* magazine, "Supertech Has Met Its Match," Jessie Hampel, February 2009.)

Now let me share some insights from some of the top IT executives in the country. Robert Willett, CEO and CIO of Best Buy (and note his dual titles!), was interviewed by Mark Hall, general manager and executive counsel for *CIO* (as in "Chief Information Officer") magazine in the April 2009 issue. Willett said, "IT needs to be more a part of the overall business. They need to know the business. Technology alone is no longer the answer to all its competitive problems." He goes on to point out that Best Buy is moving toward the "service mode" and identifies the ability to "*listen, listen, listen*" as the critical skill for the twenty-first century.

REALITY CHECK: *Is your IT team famous for its ability to listen?*

Perhaps you were thinking that these concerns affect only the private sector? Think again. Steven Warren, Deputy CIO of the second-largest federal agency, the Veterans' Affairs Administration, was interviewed as well.. He emphasized communication and empathy(!) as the keys for success in IT today, and asked his own IT people to begin seeing themselves as "change agents" within the federal government.

REALITY CHECK: *Are the members of your IT team well-regarded right now for their ability to show empathy to end-users? How about their skill as "change agents" within your organization?*

In yet another interview,, Kenric Anderburg, CIO of Phillips Health Care, observed that, today, "Managers have to move from behind their desk and visit their business units ... (they) need to understand how we work with virtual teams ... One needs to manage with the job skills of a consultant and have a deep customer service focus." He concludes: "Leave your IT hat behind—and get broad-based business knowledge. Learn the business process."

REALITY CHECK: *Does your IT team currently win praise for its "outstanding customer service"? How about its knowledge of the company's larger business processes?*

Here's the point: What it takes to succeed in IT today is different from what it took to succeed yesterday. Today, you must have a consultant's ability to build close client relationships with both internal and external customers. You must have above-average communication and listening abilities—and these are, let's face it, not skills for which most IT teams are widely praised. As if managing those transitions weren't enough, you must also realize that, when you show up for work each day, you are, whether you like it or not, operating in a *marketing* role—and you must have a clear understanding of all the advantages, disadvantages, challenges, and opportunities of that role.

Whether IT clients are internal or external, they no longer want a simple technical solution from us. In response to massive shifts in the economy, brutal competition, increased government regulations, a new era of environmental awareness in both business and government, and a host of other challenges, our clients need us to communicate effectively enough with them to deliver collaborative *business* solutions that make sense from *their* perspective—not ours.

Most IT teams are not doing that at present. If that's not a crisis, I don't know what is.

The irony here is that we are the very people who are best positioned to help the organization meet its daunting competitive and organizational challenges! We're the ones who can help improve relationships with end-users! We're the ones who make solutions possible! We're the ones who can turn bad experiences into good ones! We're the ones who should be making the case for change!

To make that case—and I believe it is a case we must make—we first must change the way we ourselves look at what we do for a living.

Recently, I came across a message from a CIO who has a keen appreciation of these issues: Deane Morrison, Chief Information Officer at Concord Hospital in Concord, New Hampshire. I've had the honor of working with Deane, and I'm reproducing his message in full, with his permission.

As my career as a CIO has evolved over many years, it has become more and more evident to me that the acceptance of an IT department within the organization it serves, its credibility with the individuals it supports, and its ultimate position and success within the organization it represents are indelibly intertwined with its ability to provide the highest quality of service. Yet the IT team itself does not always make it easy for senior management to prove that this service is being provided.

Too often, the IT organization rests on its laurels, emphasizing its technical acumen and assuming that, if its employees can "solve" its users' problems, then it has done its job. Unfortunately, this is far from true.

The measure of a high-quality, well-respected IT department lies not only in its ability to solve user issues and problems, but also in its ability to form trust and respect, to listen and empathize, and to communicate and deliver a sense of personal and organizational accountability to those same individuals.

At Concord Hospital, we have embarked on a long-term service strategy of ongoing education for our IT staff regarding the principles of meeting or exceeding our clients' expectations of the support we provide. We provide formal training in service, listening, and effective communications techniques. We role-play difficult client situations with the IT team. We involve our customers and clients in our service seminars. We develop IT and customer/client service teams and discuss opportunities to improve. We measure our clients' satisfaction with ongoing surveys. We personally call every unsatisfied client to find out exactly what went wrong and how to prevent it from happening again. And we do all this over and over again, year in and year out.

As I reflect on our current economic situation and try to help our organization develop the best strategic response to it, I can't help observing that most organizations in the health care sector (and outside of it, for that matter) are not following our example. As a result, they are experiencing problems. During the very time they need to ramp up their ability to provide good service—after all, most of us in health care have less and less capital to buy the new things our clients want—many IT organizations are cutting back on their commitment to service and to service education. Some are even outsourcing IT functions, with unfortunate results for everyone.

When the tide turns ... and it will; we just don't know how long it will take ... I believe that it is those organizations and teams that have remained steadfast in their commitment to service that will emerge with enhanced credibility, enduring respect, and a bright future. Those that skimp on service may well leave tarnished reputations that could be difficult, if not impossible, to turn around.

My advice to my colleagues both inside and outside of health care is simple: if you have to cut back, don't short-change service. Use IT to improve service levels. When the economic cloud lifts, you will be glad you did.

Deane Morrison
CIO
Concord Hospital

Deane is calling on all of us to adopt a different way of looking at information technology. Most IT teams, I find, are not making the case as effectively as he is. They are hooked on "business as usual"—and are, I believe, courting disaster.

Two paths are now open to us as IT professionals.

The First Path. If we continue to focus exclusively on interesting technical problems, if we never concern ourselves with the business of communicating with and satisfying the people who *own* the problems, then disaster awaits us and our organizations.

The Second Path. If we begin, at long last, to pay attention to the "human side" of the equation … if we look for ways to do what we do each day just a little differently … if we make some simple adjustments that allow us to deliver both high value and the *perception* of value that we deserve … then our future, and the future of our organizations, will be bright.

If the second path is the one you are most interested in traveling, turn the page. I'm going your way.

• •

A MAP OF THIS BOOK

The **Introduction** tells you why we need to bother with this discussion of IT and service in the first place.

Chapter One: Service Is Key (But Which Door Will It Open?) shows you why just repeating familiar slogans about service is no longer enough.

Chapter Two: My Passion answers the question, "If service is a universal concern, why focus on the IT team?"

Chapter Three: The Perfect Combination explains why IT and great service really do go together.

Chapter Four: What Is Great Service ... and Why Should We Care? gives you a working definition of the (overused) phrase "good service."

Chapter Five: The "Moment of Truth" Standard gives you a close-up of what makes both "good service" and "poor service" possible.

Chapter Six: Attitude 101 introduces you to the number one tool you need to turn around a botched moment of truth.

Chapter Seven: Three Points explains "points of contact," "points of development," and "points of completion" in our relationship with the client.

Chapter Eight: The First Thing Clients Want From Us clues you in to the number one priority of the people we serve.

Chapter Nine: What Else Do Customers Want? Defines what is on our list of expected deliverables.

Chapter Ten: Dirt Foot and Listen explains the two big ideas that support our entire service philosophy.

Chapter Eleven: What We May Be Saying When We're "Not Saying Anything" shows how easy it is to send the wrong message—and what to do about it.

Chapter Twelve: Active Listening Takes Practice shows you how active listening is different from passive listening … and how to actively listen while communicating with clients.

Chapter Thirteen: Building Our Service Strategy shows you the five essential steps for building this essential guiding plan for our business.

Chapter Fourteen: Building Your Client Relationship … One Client at a Time shows you how to establish and maintain rapport with individual clients.

Chapter Fifteen: Supporting the Team gives you the formula that makes it easy for everyone on our team to do the right thing at the right time.

Chapter Sixteen: Real-Life Service Success Stories gives you real-world examples of service-driven exchanges with customers.

LET'S BEGIN

LET'S PRETEND YOU AND I work together, and we're beginning a late-night conversation, in your office, about the way our company's IT department is *perceived* by the people we serve.

You're an experienced IT professional. I'm an experienced IT professional. Neither of us is particularly happy about the way people talk about, and e-mail each other about, what they think our department does (or doesn't do) all day long. So we decide to do something about it. We stay late one night. We order some Chinese food. We open our notebooks and our minds. And as the night goes on, we end up having a really great discussion with each other about … service.

Specifically, we have a great conversation about our IT department's capacity to **improve its perceived value by providing truly excellent service to our client base.**

You and I talk for three straight hours about what good service really is, how to build a service strategy, and how to create and support a service-oriented IT organization—an organization that will establish our department—and us as individuals—as truly *valued* contributors to the organization. We're having this discussion, on our own time, because we want to improve our department's reputation among all the internal clients it serves, from the receptionist to the CEO of the organization … and because we are really very proud of the value we know we can add to those internal customers.

What would that conversation sound like? What great ideas and strategies and follow-through would we come up with? What would we *do* as a result of that conversation? How could we transform the level of service we actually deliver … and the way we were perceived by the people who decide how big our budget is … or even whether we get a budget at all?

> *"No pessimist ever discovered the secrets of the stars, or sailed to an uncharted land, or opened a new heaven to the human spirit." Helen Keller*

We in the IT profession have for many years provided what I believe to be a very valuable service to our organization. But even given our accomplishments, we seem to have to work extra-hard to get the recognition we deserve.

Many times, our value is challenged at budget renewal time. We are often questioned in great detail as to why this or that expense is really required. I've experienced this phenomenon myself for decades now, both as an IT manager and as a consultant to many IT organizations, and I've come to the conclusion that this "do-we-really-need-to-spend-money-on-that" conversation is *not* the result of higher management looking for new ways to make our lives difficult. That's the working assumption that a lot of IT professionals seem to use, though: "Management has nothing better to do than to push hard on us and make our jobs more difficult."

Perhaps we don't *really* believe that—but, as I say, we sometimes act as though we do. Obviously, management has many more important things to deal with than pushing hard on IT. What management is really asking us for is the answer to these questions:

Why should I approve this request over the requests of other departments that seem to be doing a better job of returning value for the investments they're asking for?

What does this seemingly exorbitant investment actually do to enhance the company's bottom line?

It seems a little odd for us to have to face such questions, doesn't it? After all, our department has been around a while and has served many, many people. Our technology is everywhere. Most, and perhaps all, of the company's employees use our technology to fulfill their daily work requirements.

Why, then, do we feel like we are still introducing ourselves, like we're somehow the new kids on the block?

Here's why. When you take a closer look at the many components that make up the functioning body of our company, and then compare us to departments like accounting, engineering, personnel, drafting, research, marketing, sales, manufacturing, or maintenance—we actually *are* the new kids on the block!

If you looked at the typical organizational chart from the 1940s, you would have found all of those departments, or at least their close cousins—but you certainly wouldn't have found an IT department.

Old habits, and old ways of looking at the workplace, die hard. Many people still don't think of IT as part of the "standard," value-producing portion of the company. They think of us as an added, and perhaps unnecessary, expense. If we choose to do nothing about that—if we choose to focus on our resentment on the "do-we-really-have-to-spend-money" conversation at budget time, or, even worse, if we choose to pick out technical flaws in what management says or implies about our department—we will lose our customer. And when that customer is a senior executive, we shouldn't be surprised if "losing the customer" translates to "watching most or all of what we do get outsourced." That's what happens when there's a conflict in the way *value* is perceived between the customer and the vendor: The customer changes vendors!

That night, as you and I talk long into the night over our Chinese food, you and I realize that we have no alternative. We must close the value gap that our customers perceive.

So: What do we have to *do* to close that gap? What do we do to get not only the budget and the recognition we feel we deserve, but also the respect we know we have earned as a major contributor in our company's successes?

The answer is simple: We have to deliver truly superior *service* to our clients.

This is the very best place to start, because by servicing our clients well, we can also teach them our capabilities. By servicing our clients well, we can become more than a simple technology supplier: we can show our clients how we can help them function better with most of their business concerns—and improve their bottom line by helping them manage their business better.

By servicing our clients well, we can move closer to the goal of running the department as though it were a self-supporting business. And by the way, that's *really* what the higher-ups want to know at budget time:

> *Are you running this department the way it would be run if it had to operate as a stand-alone business?*
>
> The purpose of this book is to enable you to answer that question with a single, powerful word: *"Yes!"*

Pass the *moo goo gai pan,* please.

SERVICE IS KEY!
(BUT WHICH DOOR WILL IT OPEN?)

CHAPTER ONE

AT ONE POINT OR ANOTHER in our lives, we've all heard the following message, or some variation on it:

Service is key!

I've got good news. Most people genuinely believe this, or whatever version of it they've been exposed to, and also believe, at least on an intellectual level, that they should speak and act with customers accordingly.

That certainly goes for you and me. We "already know" that **service is key.** We know that, if we want success in any customer-related transaction, especially when we need the customer to continue an ongoing relationship with us, we are "supposed to" operate under this principle: **service is key!** And usually (we tell ourselves), we really do try to operate that way. That goes for the majority of people working in IT, and, I think, the majority of people working throughout the organization. It's quite rare to run into someone who actually believes that good service is *not* an important goal, and that service is *not* key for success.

There's a reason we feel that way. We are quite experienced in recognizing the below-average side of service, since we've run into the "bad service" phenomenon directly, as consumers, in any number of settings. We can all tell stories about the bad service we encountered at this or that restaurant, or while trying to register our automobile, or while trying to flag the attention of an "associate" at the local mega-sized hardware store. In fact, I'm sure each of us could come up with at least twenty unacceptable customer service experiences from our own personal data bank without any problem at all.

How often, though, do we step back from our role as "customer" and ask ourselves *what caused* the bad service episode we experienced?

THE TURNING POINT

What caused the bad service? Asking that question is the real turning point. It's the twist in the story where we become part of the solution, as opposed to part of the problem. This is the part where we start analyzing the situation to try to figure out the *reason* we were on the receiving end of an unacceptable service experience.

After all, anyone can *complain* about a waitress's surly attitude, or the surrealistically long line at the automobile registry, or the vanishing act that the person at the hardware store pulls when he spots a customer at the far end of the aisle who looks like he needs help finding something. Complaining about things is the easy part. *Making sure we don't inadvertently model the same behavior is the hard part.*

Are you ready to encounter the turning point for yourself? Great.

Think of a particularly bad customer service experience right now. Call a specific, memorable incident of terrible service to mind and, once you have identified a particular event, turn the page and give your best answer to the questions you see there.

- What is the first thing you remember about the experience? What was it specifically that *caused* you to think of that episode as a bad service experience? What *happened to you* as a result of the bad customer service experience?
- How, specifically, did the experience affect your day? Were you inconvenienced in some way? Did you have to spend extra money? Did you waste time? Were you frustrated?

WHAT *REASON* DID YOU COME UP WITH?

Maybe the *reason* that came to your mind was something like the other person's rudeness or lateness. Or maybe it was the fact that you were rushed or ignored, or spent more money than you thought you should have spent, or had to a listen to a rude remark the person made.

That's all good information. Still, I'll bet you have also had bad service experiences when *nothing* was said, or when you never actually encountered *any* unpleasant actions you could put your finger on, yet you still left the premises with a negative experience ... or maybe even said to yourself, "I will never come back here again." Has that ever happened to you? Try to think of a specific instance like this, too. Do so before you turn the page.

Most of us have had a "bad service" experience of this kind at one point or another in our lives. If you are one of the few who haven't, let me be the first to congratulate you!

My estimate is that roughly nineteen out of twenty people can easily come up with a "bad service" experience where the problem was the chemistry between you and the other person. That bad chemistry simply poisoned your experience with the company, even though there was no clear word or deed you could have complained about.

For those of you who *have* had this experience—nothing technically wrong was said or done, but you still left the counter with a funny feeling that said to your inner core, "I am not at all satisfied here"—I have a question. *What in the world do you think was going on during those experiences?*

Let's see if we can come to understand what happened to cause you to feel this way. Let's try to identify exactly what takes place in this very common scenario. Follow me closely through the next set of ideas: once we get our heads around this next part I think we will begin to understand just how difficult it can be for us—or anyone—to follow through on what we think we "already know" about customer service—namely, **Service Is Key.**

Did you think for just a moment about what exactly *caused* you to have this feeling of poor service? Again, we're talking about an episode where nothing actually "went wrong" in the interaction that you had with the service provider—but you nevertheless came away from the exchange thinking something like the following: "Man, that was awkward." Or: "I can't put my finger on it, but something about that person just rubbed me the wrong way." Or even: "Something about this place gives me the creeps—I'm never coming back here again."

For this kind of exchange, and we've all had them, we often can trace the bad feeling back to a person I describe as having a DILLIGAD attitude. That's an acronym for

Do

I

Look

Like

I

Give

A

Damn?

Haven't we all met more than our fair share of DILLIGADS? The DILLIGAD is the person who can make your skin crawl just by looking at you, or not looking at you, as the case may be. This is the person who can make you wish you'd wound up with someone else to talk to—*just by the way he or she said a single word or phrase,* like "Next!" or "Can I help you?"

It's easy to figure out why you felt bad about the service you received when someone serves you food that's cold, or leaves you standing for much longer than you should have been standing, or made you explain something that they should have already known about your situation. But when someone fixes you with a look that says, "Do I look like I give a damn?" *everything else can go "right," and your relationship with the company providing service can still be all "wrong!"*

11

Sometimes, the DILLIGAD moment is a much more significant problem for us than an actual lapse in a deliverable we're supposed to receive from the organization! In fact, even if this person provides an outcome that we're actually looking for, even if everything comes in on time, under budget, and meeting all the specifications, there can still be a big problem if there's a DILLIGAD attitude to deal with. We may walk away from the exchange convinced—at least an emotional level—that we received terrible service, even though we technically "got what we wanted"!

I will go into much more in detail about this DILLIGAD phenomenon a little later in the book. What I want you to consider now is this downright terrifying prospect:

Maybe we have *all* dealt out a little DILLIGAD in our careers ... without even knowing it. The sad fact is, it is far easier than any of us would like to admit to fall into a DILLIGAD attitude, without our having any idea that that's what we're doing.

This unpleasant fact is one of the big reasons it can be difficult to follow through on what we "already know" about customer service, that *service is key*. Yes, service is key ... but DILLIGAD service opens a door that neither we nor our customers really want to go through!

BIG QUESTION

ARE YOU RUNNING THE I.T. TEAM THE WAY IT WOULD BE RUN IF IT WERE A *STAND-ALONE BUSINESS?*

MY PASSION

CHAPTER TWO

YOU MIGHT WELL ASK: Since service is a universal concern, why am I singling out the information technology department?

The answer is simple. I began my career in IT way back in the early 1960s, and service within that part of the organization is my passion. I rose to assume a number of different management titles within the profession, rising eventually to the level of Corporate Director of Information Services. I have overseen both large and small operations and have, over the years, worked out many service issues that connected not only to staff, management, vendors, and consultants, but also to both internal and external operations and to larger issues like public relations. Over the years, I have worked for such respected companies as Raytheon (defense); The Hoechst Corporation, International, (plastics and pharmaceuticals); and Public Service Co. of New Hampshire (utilities), among other companies.

In 1984, I founded a consulting firm for the sole purpose of improving IT's client interactions. I provided a variety of offerings, and I defined our mission as follows: *a dedication to the development of the human side of technologist.*

I'm proud to be able to say that I did well with that consulting firm; I sold it in the year 1999. During the twenty years I spent as a consultant and trainer, I worked with IT organizations throughout the

U.S., Canada, Europe, and South Africa; before founding my consulting firm, I earned my degree in computer science from Northeastern University in 1973. I am also proud to say that I attained the highest earned recognition presented by the National Speakers Association, that of Certified Public Speaker, in 1998. I have spoken as a keynoter to IT people at such organizations as the Chief Information Officers' Conference, the International Technology Conference (London), the Data Processing Management Association, the Edison Electric Institute, the American Gas Association, Industry Week Conventions, and many other groups.

Now, I don't tell you any of this to impress you, but rather to impress upon you that my passion really is with the information technologists of this world. That's because I am one of those information technologists—and I'm proud of it!

I have come to know first-hand the great achievements that we in IT organizations have accomplished, how our profession has changed, and how our clients have become more demanding than ever. I have come to know, too, that we face some serious challenges. I see those challenges as an open door to great opportunities for us to grow and play an even greater part in the success of the organizations we serve.

I do not wear rose-colored glasses. I understand the tremendous pressures that are brought upon us, week after week, quarter after quarter, and year after year, to bring the right technologies to the right place, for the right reason, at the right time, for the right price. Of course, many of the people we deal with do not have the understanding of IT on a personal level that we have. Their understanding of our real technical capabilities, in many cases, is well-intended but sadly limited. Bridging that gap on a personal level is part of our job.

What I want you to know is that I have worked with many successful IT departments in many different industries. I have worked with teams that faced major challenges, and I showed them how to resolve those challenges and **win the recognition they deserved as valued contributors to the mission of their larger organizations.**

My passion is helping such teams to focus on *service* as the key to *earning* the trust—and the business—of their internal customers.

In this book, I want to show you exactly what I showed them, because I believe you, too, deserve to achieve this level of recognition within your organization, and that you can achieve it based on one thing and one thing alone: your ability to properly service your client needs. And this brings us back to a critical point: *service is key*.

The right service key will open a new way of conducting business for you and your entire team. The right service key will take time and effort to learn, but it will open the right doors for you—if you are willing to move past "business as usual" so you and your team can move toward being the kinds of IT professionals who don't have to say, at the end of the day, "I don't get no respect." (Or, even worse, as Rodney Dangerfield might say: "I ain't got no budget.")

Within this book, you and I will have a conversation about how we can work together to develop and clearly understand the power of a new *vision,* a new *attitude,* a new *approach,* and a new *personal conduct* in the area of information technology service. In short—we will find the right *service key,* the one that opens the right door. At the same time, we will learn the basics of a brand new *management philosophy,* a philosophy that will guide us towards the *respect*—and, yes, the *stability*—we deserve on both the individual and the team level. The right *service key* really can make that kind of difference.

Let's take a closer look at our understanding of how we operate as an IT team by filling in some of the blanks.

LET'S ASK OURSELVES, "WHAT REALLY IS SERVICE?"

Fill in the blank lines

Why would great IT service be good for IT?

_____
_____
_____

What might the pitfalls be?

_____
_____
_____

How do we set measurements? What should they be?

_____
_____
_____

Is there one service model that fits all?

_____
_____
_____

Guarantee satisfaction—can we do that?

_____
_____
_____

What do we do if we fail?

_____
_____
_____

What would be the acceptable error rate?

_____
_____
_____

THE FORMATION OF AN IT SERVICE STRATEGY

Guidelines for Your Review

What are the rules by which we operate?

_____
_____
_____

Are they service-oriented?

_____
_____
_____

Clearly measure management expectations within and outside of IT.

_____
_____
_____

What do we accept as acceptable service?

_____
_____
_____

Do we want "all" IT business?

_____
_____
_____

Promote the value of IT Service.

_____
_____
_____

WHAT IS OUR IT GENERALLY APPLIED SERVICE MODE?

Reviewing our current status of client interface, do we:

1. Deliver technology as specified by
   our clients?   Yes _____   No _____

2. Truly understand our clients' needs?   Yes _____   No _____

3. Add value to our clients long-range
   business goals through technology?   Yes _____   No _____

4. React to client requests promptly?   Yes _____   No _____

5. Contribute to our clients' business
   problem areas?   Yes _____   No _____

6. Also seek clients' unmet requests?   Yes _____   No _____

WHAT IS OUR CURRENT MODE OF OPERATION?

Are we built on speed of activity
and data retrieval?   Yes____ No____

Is speed for transactions and connectivty
our main objective?   Yes____No____

Is our focus based on keeping
our operating cost low?   Yes____ No____

Do we make it convenient to access our services?   Yes____ No____

Do we limit ourselves to meeting
client expectations?   Yes____ No____

Do we allow for increases to our
expenditure to satisfy our clients' needs?   Yes____ No____

Do we strive to build strong client relationships?   Yes____ No____

Is our main objective to exceed
our client's expectations?   Yes____ No____

Are we prepared to add additional expenses
to our budget in an effort to provide
a premium-level service?   Yes____ No____

KEY TERM

DILLIGAD = *DO I LOOK LIKE I GIVE A DAMN?*

THE PERFECT COMBINATION

CHAPTER THREE

CONTRARY TO WHAT YOU MAY hear, IT and great service really do make the perfect combination.

In fact, when we put those two things together, IT and service, we lay the foundation for something truly wonderful: *success*. (That's another passion of mine, by the way: succeeding!)

Here's one of the reasons I feel so strongly about this: Our value goes largely under-noticed, and that's unacceptable. After four decades in this field, I still feel that the contributions IT makes are not well understood by the many, many people who benefit from what we do. And that means we have some work to do when it comes to communicating our value. *The fact that our value is under-noticed is our own fault ... and our own very special opportunity.*

We can use service to do a better job of spotlighting our value. And we must!

Over the years, in training programs, I've asked countless IT teams this question: "How many of you feel that you are truly part of the *fabric* of this company that you work for?" Inevitably, a few hands go up—but most hands stay down.

I then ask, "How many of you feel more like you are a piece of *lint* on the fabric of this organization ... and someone is always trying to brush you off?" The hands always *shoot* up in the air, and everyone shouts, right out loud, *"Yes!"*

Sad but true: That's how most of us IT professionals feel. Like a piece of lint! And I am here to tell you that the "lint factor" is our *own damn fault* ... because we have not done the job we should have done, and could have done, when it comes to servicing our customer and communicating our value.

Going a bit further, I have also asked many senior IT managers how well they feel their IT organizations are *understood* at the highest levels of the company. Far too many of these managers feel that their team's potential to enhance the bottom line is simply overlooked. Too often, IT is seen simply as a "must have"—but as one that comes at a heavy cost in terms of corporate overhead.

There are many possible reasons for this misunderstanding of our true value. From the very earliest days of IT—and by that, I mean the period when I launched my own career—there were prolonged periods of "down time" and countless project estimates that turned out to be inaccurate. I am here not only to tell you that these problems were worse in the old days than they are today, but also to acknowledge that those of us who blazed the trail back in the old days were the ones who helped to make your life difficult today. We were the ones who established IT's less-than-stellar reputation when it comes to things like down time, scheduling errors, technical capacity, and good, old-fashioned human error. In other words, when you showed up on the job, you probably noticed that people were already jaded about these things, which didn't make your life any easier or make it any easier for you to solve the problems you faced. I am here to confess that we who went before you were the ones who *made people feel jaded about these issues in the first place,* and I, for one, am sorry about that. Many people are still skeptical about IT, and they are skeptical, in large measure, not because of anything that you have done, but because they remember the old days. The jokes and stereotypes associated with IT became hardened perceptions, perceptions that our clients have seen or experienced and passed on to others. Please keep in mind that *perception is reality* to anyone who has formed a certain perception.

That's one problem. Another has to do with the messages we have chosen to send, both in the old days and today. Many of us in

IT probably have worked a little too hard to *establish* ourselves as outsiders. Whether we positioned ourselves as the "nerd" or as the "high priest," we were often sending a message to our clients that we speak a different language than they do, that we live in a world that they don't understand and couldn't possibly understand, or even that we are aloof and sometimes have no patience or interest in anyone who doesn't understand our technology or our language. Over time, we have built up a unique mystique about ourselves and our technology ... and perhaps even enjoyed that a bit.

The time for us to maintain that mystique—if, indeed it ever made sense to maintain it—has passed. Times have changed, and they have changed drastically. Our clients are now much more familiar and knowledgeable about many aspects of technology, and they are in a better position than they ever were to make good decisions about what type of technology they want to use and what kind of software will best fit their needs. The playing field is now a little closer to being level..

This greater familiarity and comfort with technology has brought both advantages and challenges. People have more access to information than ever before. The potential improvements in productivity and innovation, and the opportunities of whole new media and industries within incredibly short time-spans, are all breathtaking. At the same time, though, issues such as the management and security of essential corporate data, the privacy of employees, the runaway cost of technology purchases, the correct placement of these technologies, the identification of the right data pathways, and the risks to the larger community connected to the possible manipulation and storage of vital data all give cause for concern.

It would take a very different kind of book to evaluate all those opportunities and all those challenges, probably a very thick book that most people wouldn't finish. What I want you to understand here and now, though, is that, because of these changes, we have reached a point in our history where IT must be seen not only as a technology supplier, but also as an integral contributor to the essential job of managing our corporation's daily business. To give only one obvious example, it is now beyond dispute that we must help our companies to be more "data safe"—and we cannot really do that job well if we are held to unhelpful stereotypes or regarded as outsiders, as people who do not live in the "real business world" of costs and benefits that everyone else lives in. We do live in that world, and we must make sure people know that.

Our organizations *need us* to be part of the fabric of the organization. One way to get us there, one way for us to reposition ourselves and change our misperceptions by our clients, is to not only provide the great work that we already do provide, but also to provide it with a service *attitude* that redefines how internal clients think about IT.

> **Improving our service attitude is not a silver bullet. It won't fix everything. It can, however, move us forward toward the recognized and fully respected role within the organization that you and I both know we deserve.**

NEW NEW NEW

NEW VISION + NEW ATTITUDE + NEW APPROACH + NEW PERSONAL CONDUCT
= *NEW OPERATING PHILOSOPHY*

WHAT IS GOOD SERVICE ... AND WHY SHOULD WE CARE?

CHAPTER FOUR

AFTER SPENDING DECADES IN THE world of IT, I have reached the sad but inevitable conclusion that most of us don't really know what good service is.

We appear to recognize *bad* service when it happens to us, but we have little to add to the discussion when the time comes to identify specifically what *made* an interaction a good or bad service experience. When pressed, some of us are likely to explain that it is simply "the person's attitude." Others of us will say that it was the length of time that it took to complete the transaction. And others of us will insist that the problem had something to do with the lights, the music, the kind of uniform the person was wearing, or even the person's hairstyle.

Well—what was it that was missing? What *is* good service, exactly?

We can say that we each have our own individual definition, but if we really plan to make our organization "service-oriented," it behooves us to have an answer that's a little more quantifiable than "I know it when I see it."

We all agree, I think, that we want good service to be an integral part of our daily offering to our clients. If that's the case, we need a much better understanding of the meaning of "good service" … and we also need a better understanding of how we can standardize this "good service" experience—whatever it actually is—and make sure our whole team is delivering it consistently over a sustained period of time.

Are you like most of the IT professionals with whom I have worked? Are you not *quite* sure how you would define "good service" for a new team member who asked you how he could tell for sure that he was delivering it, day after day?

If your answer to this vitally important question is "yes," let me ask you to consider the following definition, which I have used for years and am happy to stand by until something better comes along:

Good service is leaving a good impression with the customer during a Moment of Truth (MOT).

Now, I'd like to be able to tell you that I came up with the phrase *moment of truth,* but I didn't. That very-nearly-self-explanatory phrase (don't worry, I'll fill in the blanks in a minute) was the brainchild of one Jan Carlson. He didn't work in IT, but he shed so much light on the question of customer service that I'd like you to consider him, for the moment, an honorary member of the fraternity.

Carlson took over as President of Scandinavian Airlines Systems (SAS) when that outfit was very close to bankruptcy. He turned his airline around by explaining to his employees, in various compelling ways, that service is determined by seemingly tiny chunks of customer experience that happen at critical moments—moments of truth. These moments of truth, Carlson explained, end up having a huge impact, either positive or negative, on the perceptions of the only person who ultimately gets to decide whether the experience was "good" or "bad." Of course, that's the customer.

Carlson's theory that there were some moments that were far more important than others in determining the customer experience had a huge impact on the field of customer service.

Let's take a look now at how these moments might play out in your real world—as a customer of, let's say, a gourmet restaurant.

You and your significant other are out for a very special evening together. You have selected the best restaurant in town. You phoned ahead and placed a reservation for the perfect time. You are sitting there together at a nice table with a beautiful view of the ocean, and you're gazing into each other's eyes with complete attention, whispering beautiful declarations of love to one another. Life is good.

Ten minutes later, the waves are still churning, and you're still gazing at each other lovingly, but the words of love aren't coming quite so easily. Your water hasn't shown up yet. Life is okay.

Fifteen minutes later, no waiter has materialized. You try to gaze at each other with complete adoration, and you try to make small talk, but it takes a bit of an effort now. You're hungry. Life is getting less okay by the minute.

Twenty minutes later, you still like the look of your partner's eyes well enough, and that big old ocean is fine, but you just can't believe that you haven't been able to attract a waiter to your table. Not only that, you can't seem to catch the eye of any kind of employee in the establishment. Everyone seems to have vanished.

Finally, you spot a waiter walking near your table. Relieved, you say, "Hi there—may we have some service, please?"

The waiter smiles and replies, "I'm sorry—I don't work this section." And walks away.

What would you do? You may feel a little frustrated with that answer and may wonder whether coming here in the first place was all that good an idea.

Some people, I'm sure, would go back to looking adoringly at their partners. Others, though—maybe a majority—would say, "I'm going to go find us a waiter." And *some* people—not a lot, but some—would say, "Let's get out of here. That waiter had no right to talk to us that way, and I don't need this type of bad service."

How many moments of truth were there in this scenario? I count at least two. The first was the initial time you spent at the table. That was an opportunity for the restaurant to impress you by getting a waiter to your table shortly after you sat down. Unfortunately, the restaurant missed that opportunity. And the second moment of truth, of course, was when the waiter gave you the brush-off. The restaurant missed that opportunity, too. As of now, the restaurant is zero-for-two in the moments of truth batting race.

Good service *means taking advantage of moments of truth in a way that allows you to leave customers with a positive impression.*

Poor service *means missing the opportunity to capitalize on a moment of truth and leaving customers with a negative impression.*

GOOD SERVICE

... MEANS TAKING ADVANTAGE OF *MOMENTS OF TRUTH* IN A WAY THAT ALLOWS YOU TO LEAVE A POSITIVE IMPRESSION WITH A CUSTOMER.

THE "MOMENT OF TRUTH" STANDARD

CHAPTER FIVE

LOOK AT THAT SITUATION AGAIN: We haven't even ordered, we haven't even met our real waiter, and yet our antennae are already picking up "bad service" signals. That simple example is realistic enough, isn't it? Hasn't something like that happened to you—perhaps in the recent past? What made it happen? A botched *moment of truth!*

Moments of truth are what make the difference in terms of customer perception. Are they well-handled ... or fumbled? In that restaurant setting, the waiter's ability to arrive at our table promptly and the kind of response we get when we ask for service each count as *moments of truth* that could have been, and should have been, identified ahead of time. In fact, each could have been *strategized* better ... ahead of time.

This is how "good service" and "bad service" happen. Each of us, in our own analysis, creates a unique reaction to a *moment of truth,* which can usually be identified ahead of time by the service provider. At different times during the service cycle, we may be more or less likely to overlook a problem we experience as customers during a *moment of truth.* For instance, if you have been a regular at this restaurant for years and you know that four of your five favorite waiters are all either out on vacation or out with the flu, you may be inclined to come back again—even though you had a couple of blown MOTs during one visit.

All of us, whether we are in IT or not, assess *moments of truth* on our own terms—as customers. Our judgments come from all aspects of our interactions: how long we have to wait (or don't have to wait), what is said (or not said), the tone the person takes, the pitch of the person's voice, the sounds we hear around us, the things we see, the things we feel, the things we smell, even how we're feeling emotionally that day—all of these combine to determine how much "slack" we're willing to cut the organization (if, indeed, we're willing to cut it any "slack" at all) when we show up as a customer. Our reactions to those *moments of truth* are not only what determine how we feel about that particular exchange—they can determine how we feel about the overall quality of service that we have received and how we feel about the company as a whole.

Here's the good news: we really can determine your customer's most important *moments of truth* ahead of time. Now here's the sobering news: the way we "deliver the goods" during those moments of truth will determine how our (internal) customer experiences interactions with the IT department.

What happens during these critical moments will, over time, determine whether we are perceived as delivering "good service" or "bad service."

Now, as a member of your IT team, use the following form to take a stab at a big question: How do you think your customers perceive what you do?

OUR SERVICE LEVELS AS SEEN BY OUR CLIENTS

Moments of Truth

What exactly is meant by the term "moment of truth"? Based on what you now know, write your best definition below:

_____

_____

_____

_____

_____

_____

_____

_____

_____

_____

_____

_____

_____

_____

_____

_____

_____

_____

_____

_____

_____

_____

After that definition of service, you may be wondering, "Can we really live up to these critical moments, considering that there are so many potentially negative reactions to them? Can we really expect to compete within that environment?"

The answer is a resounding *yes!* We certainly can. But in order to compete, we must examine what we are already using as benchmarks for "good service" in our interactions with internal customers and see how close they come to the *moments of truth* model.

I have asked many IT professionals over the years exactly what it takes to be a service-oriented organization. Typically, I get answers like this: "To me, to be a service-oriented IT organization, we must give our customers what they want, when they want it."

It sounds good, but is it actually possible? Can you really give them exactly what they want, exactly when they want it? Is that what they're really expecting from us? Is it a realistic goal? How long could you do that? What impact would trying to do that for everyone have on your resources and your schedule? How long could anybody, in any business, including the restaurant business, expect to give the customers what they want, when they want it, 100 percent of the time?

You get the idea. Simply saying "yes" to any given client request, on the schedule the client suggests, is not the right strategy. So what's a better idea?

What we must do is start building a service *mentality*, and we must simultaneously find out about the basics of our service requirements—the "make or break" elements that are likely to turn a *moment of truth* into a "bad service" perception. We must develop a clearly-understood process that all staff members can embrace when it comes to executing those "make or break" elements.

To help us understand this service process, its components, and how it affects our clients, we will assume that we are working at a fictitious company, Innovative Technological Systems. You'll be reading more about this company a little later on in this book. For now, let's take a look at *why* our clients may not be sending us all the signals we'd hope that they would about poor service. Take a look at the following form.

Why don't clients talk to IT about their bad service experiences?

- They believe talking is simply not worth their time.

- They believe that no one really cares.

- They believe that no one will take corrective measures.

- There is simply no good medium for expressing their concerns.

***Guess what?* They will talk to everyone else!**

WHY OUR CLIENTS MAY NOT TALK TO IT ABOUT OUR LESS-THAN-EXPECTED SERVICE LEVELS

Reasons

Reasons given by our clients:

_____
_____
_____
_____
_____
_____
_____
_____
_____
_____

Reasons given by IT:

_____
_____
_____
_____
_____
_____
_____
_____
_____

OUR OPPORTUNITY

WHAT HAPPENS DURING *MOMENTS OF TRUTH* DETERMINE WHETHER OUR TEAM IS PERCEIVED AS DELIVERING "GOOD SERVICE" OR "BAD SERVICE"

ATTITUDE 101

CHAPTER SIX

THERE IS *ONE THING THAT* can, in my experience, transform even a botched moment of truth into a positive experience—into a reason to say, "Hey, there was a little problem up front, but you know what? I really liked the service I got at that place."

This is something that might just change your opinion as a customer about whether you've had "good service" or "bad service." Even if you've been sitting at the restaurant waiting for far too long, even if the server you asked for help told you that it wasn't his table and kept right on walking, if someone at the restaurant found a way to put this asset to work before you actually left the premises, you might actually conclude that you'd had a good experience!

What is it?

One powerful word describes the single, simple factor that can turn around even a blown moment of truth: *attitude.*

Attitude has a *huge* impact on our ability to render great service. This is one of those issues that we often believe others need to work on, but I can assure you that everyone who interacts with internal or external customers—and that includes all of us in IT—has the responsibility to study and improve the attitude we display during our exchanges with customers.

Because attitude has such an immediate impact on the customer experience, it is the very first piece in the development of our service model. Translation: **DILLIGADS need not apply!**

If you are reading these words, you've already offered evidence to support the proposition that you usually try very, very hard to have a positive attitude when dealing with anyone and everyone—certainly when dealing with your internal clients. It's possible, though, that, like me, you have occasionally forgotten just how easy it can be to display a negative attitude. Unfortunately, in many situations, a negative attitude is the default setting for human beings. *If you don't make a conscious effort to show a positive attitude, a negative attitude can display without your even meaning for it to display.*

I know that you, like most of us, try hard to be positive in interactions with your internal and external clients. But tell the truth: have you ever slipped from that standard, as I did? Even once, during a "down day"?

If your honest answer is "no," then you have my congratulations. You are a true IT pro—and no doubt your focus now is on sharing your top-drawer skills with everyone else in the organization. Come to think of it, if your honest answer to this question is "no," you may not even need to read the rest of this book. You should probably be the one writing a book on this topic.

MEANWHILE, BACK ON PLANET EARTH ...

Now for the rest of us. You, like me, answered "yes" to this question, and that means you have a lot in common not only with me, but also with most of the IT people I work with and, in fact, with just about every other IT professional now drawing breath. Sometimes, even though we *want* to deliver a positive attitude to our clients, we end up not delivering that attitude as often as we'd like.

In other words, we're human.

As human beings, how can we expect to acquire this positive attitude and maintain it on a consistent basis? This is a completely fair question. Let's look at it together right now.

We should start by recognizing that we are each unique as individuals. We each have different backgrounds, different educations, different views, different understandings of how the world works, and different core assumptions. As human beings, we feel pretty confident that the way we see the world is the "right" way, and we are likely to want to assume that, if our way is the "right" way to look at the world, then others should be prepared to look at things in the "right" way—our way—as well.

So how can we get all these different people with their different views together, all being open to the *client's* viewpoint and all marching to the same drumming beat?

First, let me say that I consider this diversity of viewpoints to be an *asset* to our cause, and not in any way, shape, or form an *obstacle*. We all deserve to be respected for our viewpoints and our experiences. Second, I truly feel that the majority of us humans, regardless of our unique viewpoints and our experiences, have a genuine desire to do the right thing towards our fellow human beings. Wouldn't you agree?

My assumption is that you *didn't* get up this morning and say, "I will try my best to mess up with my coworkers or friends today." Most likely, none of us did that. (If you really did have that attitude when you woke up this morning, which I doubt, then I suggest that you seek professional help immediately.)

So, we are each unique as individuals, and we each have a desire to do the right thing in our interactions with others. By extension, I think I can also assume that we are all open to improving our own lives at any given moment and, as a byproduct, growing our careers. With all of these as givens, how do we start to improve our own unique client interaction *attitude?*

To do this correctly, the very first thing we must do is embrace the *moment of truth* concept we encountered a little earlier in the book—and do a thorough check of ourselves and get very, very clear about when and how we and our colleagues may, on a bad day, inadvertently mess up a specific *moment of truth*.

We'll start that process in chapter seven. Right now, though, I'd like you to consider the following sobering statistics.

Why Is Good IT Service *So Important?*

Surveys keep on showing that clients leave due to *poor service*. (Note: In our world, "leaving" may mean downsized budgets or outsourcing!)

Typical Survey results:

70% leave because of negative interaction by the business towards the client.

12% don't like the product.

7% leave for the competitor.

6% build new friendships with other providers.

5% move away or no longer use the product/services.

Do those percentages reflect us in any way? If so, what should we do?

UNLUCKY SEVEN

RESEARCH SHOWS THAT *SEVEN OUT OF EVERY TEN* CUSTOMERS WHO LEAVE US FOR THE COMPETITION DO SO BECAUSE OF SOME NEGATIVE INTERACTION WITH US.

7

THREE POINTS

CHAPTER SEVEN

WE MUST EACH, AS INDIVIDUALS, embrace the *moments of truth* concept I shared with you a little earlier in the book. We must each be prepared to do a thorough, personal DILLIGAD check on a daily basis so we can identify when we may inadvertently fumble the *moments of truth* that may come our way as individuals. For safety's sake, we should each create a personal list, one that we update regularly, of the things we learn about ourselves, on an ongoing basis—the parts of our own personality that leave us vulnerable to the DILLIGAD syndrome! We don't have to share this personal list with anyone, but we should make an effort to create it for ourselves and review it regularly.

That's on the personal level. We should also assemble a team.

INNOVATIVE TECHNOLOGICAL SYSTEMS GATHERS THE TEAM

Here at our company, Innovative Technological Systems, we need to get a representative group of our staff together, one that is well-selected to achieve a good cross-section of our staff in terms of both personality and technical expertise. We should then assign a facilitator to manage these sessions and start listing out **what a typical client has to go through each time they interact with any part of our organization.**

As a starter, I suggest that we think along the lines of the client's **point of contact** with the tech team at our hypothetical company, Innovative Technological Systems. Assume that you and I work for that company and that we have been assigned to a task force on improving service levels. Ready?

Let's take an area that most of us are very familiar with here at Innovative Technological Systems: the call center. At other companies, of course, this is also known by names such as the help desk, the application development department, or good old tech support. Whatever you call it, it's the means by which our clients initiate a request for help from us. This is where we get things started with the client, so it's obviously a critical part of the relationship to example. At Innovate Technological Systems, the call center is important because it's the place where the *point of contact* happens.

First, we appoint a facilitator for our task force. Then it's time for us all to start looking at things critically as a group. Here are the kinds of opening questions we should be asking ourselves during this meeting:

- As a general rule, do we make it clear to our clients exactly what procedures and requirements we need before we can act on a client request? Or do we make them guess what we need?
- Are our policies written clearly in understandable business English? Or are they written in our own hard-to-understand insider jargon?
- Does the client leave his or her first interaction with us knowing what's going to happen, who's responsible for making it happen, and whom he or she will be hearing from next?
- How many potential negative *moments of truth* could we be asking the client to experience before we accept responsibility for taking action on his or her request? For instance, how much fun is it for them to fill out forms that we may have created for them in order to get started?

Of course, the facilitator needs to keep a running summary of all these findings. These questions are designed to open up the process;

we'll be getting more detailed in a moment. As we approach these questions, we need to be open-minded throughout this discussion. Our team needs to get used to identifying each small step that our clients experience. Then we need to become ruthlessly honest with ourselves about how many opportunities there may be for our clients to experience a negative—or a positive—*moment of truth* as a result of those interactions.

GETTING SPECIFIC ABOUT THE POINT OF CONTACT

Now we're going to break the process down a little so we can see exactly how the discussion is likely to play out about specific elements of our service process. Let's say that one of our clients is placing a typical call to the help desk. What would you think would be the very first *moment of truth* that person would encounter with us?

Does our client have easy access to a help desk name list or departmental list that informs him or her about whom to contact or how to get access to a particular area of expertise? If not are we looking at our first possible *moment of truth* that's being mishandled? If so, what happens when the person actually picks up the phone or e-mails us? How many times does the phone have to ring before someone answers it? (One? Two? Three? More? If the person has to wait for more than three rings, you are likely to be in the *mishandled moment of truth* zone.) How long does it take someone to get a response to an e-mail message? (Thirty minutes? An hour? By the end of the day? "When we get around to it?" Again, beware of entering into the negative *moment of truth* arena.)

When a call is answered, is it a live, human voice that our clients hear—or is it a computerized reply? If it is a human voice, is it cheerful in nature? Is the person speaking clearly and understandably? Does the person who answers the phone have a heavy accent that is a challenge to understand? Are clients typically placed on hold while research for the problem is being conducted? If so, do we apologize for having to put the person on hold and give an estimate of about how long it will be? Do we return periodically to assure the client that action is in fact going on to move this request forward? Is this response time measured? Do we check in, say, every three or four minutes, or does the client simply "hang in there" for who knows how long? Is there some form of music playing in the background while the person waits, or is this simply "quiet time"?

Here's a big one, and one of the most important questions to consider:

How many steps do our clients normally have to take to get a commitment of support from us?

Compile all the observations that your group arrived at for the initial call to the help desk. Take a close look at your results. If you are a typical center, and the group has been ruthlessly honest about the answers to these questions, you will probably be looking at each other and saying, "*Wow!*" at the end of this exercise. As in:

- "Wow! I didn't realize everything they had to go though to just ask us a simple question!"

By the way, if you *don't* have that kind of a reaction, there are only two possible reasons: (a) you're already delivering superior service based on having identified the most important early *moments of truth* ... or (b) you're kidding yourself as a group.

POINTS OF CONTACT: TWO FLAVORS

As you prepare to formulate these questions and examine their answers as part of a group, let me "raise the bar" and ask you to identify two different *moments of truth* tracks:

- **Points of contact** with people your organization has never, ever made contact with before ... and ...
- **Points of contact** with people who have been down the "service road" with you before.

Even if these two groups of people reach out to your organization through identical channels (and they may not), you need to identify which is which very early on in your interactions, and you need to track them separately, as well.

Why go to the trouble of figuring out which of these two groups you're talking to? Because the guiding strategies for delivering on a moment of truth with these two constituencies are going to be very, very different!

The very first time someone comes in contact with your work group, the strategy for delivering on a *moment of truth* is to **make a great first impression.** How? By making the person the undisputed *star* of this conversation. The best way to do that, of course (after you've confirmed that this is in fact the first time the person has reached out to you), is to truly listen to the person, to ask questions relevant to his or her situation, and to focus all the energy and attention you possibly can on the goal of building up rapport with this person. Use paraphrasing to show that you really have heard the person's input. Give feedback where appropriate. Never, ever say "No," or "That's not our policy," or "We can't do that"—or any of the thousand variations on this theme—during a call with someone who is connecting with your team for the first time. Leave all the possibilities open, and see what you can do. If the answer is still "no," you can break the news—tactfully— on your second encounter. Make sure your management knows that you're dealing with a "first-timer"—there may be exceptions to some of those hard-and-fast rules we all hear about so often when there's an opportunity to win or keep a brand-new customer.

The second or subsequent time someone comes in contact with your work group, the strategy for delivering on a *moment of truth* is very different. Your goal is now to **figure out exactly what this person expects from the discussion.** Whether or not you can deliver on what the person expects is a whole different topic. Your goal going in is simply to find out *what would make the world spin more smoothly for the person who is now reaching out to your team.* Get a clear understanding of exactly what this client *expects* at this point. Take full, detailed notes. Don't try to disengage until both of you are completely clear as to what the expectations are. You, as the service provider, must now take ownership of this problem and commit to take further action. I'll be sharing strategies on how to do that later on in the book; for now, notice how different this discussion is from the one with the person we have *not* interacted with before. You are now the representative— the ambassador, if you will—for a country that this person has visited before. You must project a clear willingness to be good guide and a good advocate. That means no complaints, no DILLGAD, no guilt trips back to the client for not having provided information he or she "should have known" that you need.

> *These two groups* experience *moments of truth from totally different perspectives ... so it's not all that surprising that we should have to plan for them, and track them, differently.* **Be sure you talk about the different kinds of points of contact, the different groups, and the different strategies for interacting with them during your team meetings.**

Now that we've gotten a little momentum going at Innovative Technological Systems, we're certainly not going to stop with **points of contact.** It's time to look at two more phases of the service experience where moments of truth are waiting to be understood more clearly: **points of development** and **points of completion.**

POINTS OF DEVELOPMENT

Whether it takes us thirty seconds over the phone, ten minutes in person at a help desk, or a period of weeks or months using multiple means of contact, there is a middle period where we're actually working on the problem the client has given to us. This middle exchange is known as our **point of development.**

Just as we gathered a cross-section of the team at Innovative Technological Systems and started to ask ourselves important questions about the point of contact experience, we also need to get a better understanding of the point of development we deliver to our clients, and we need to find out where the most important moments of truth in this phase of the experience lie.

We should ask ourselves:

- What is the dividing line between the point of contact and the point of development?
- How does the client know that he or she is moving from one to the other?
- What is the timeline for completion of the point of development phase?
- How much does that timeline vary from case to case? What are the extremes on either end? What is the *likely* resolution

timeline for the most common problems we encounter? What *usually* happens?

- Who, if anyone, is currently responsible for giving clients a clear, comprehensible assessment of the problem, what we're going to do about it, the likely timeframe for its resolution, and a clear contact point the client can use in the meantime?
- While we are working through the point of development phase with the client, does the client know what's happening, what to expect, and when to expect it?
- Who is accountable to the client during the point of development phase?

Last, and perhaps most importantly:

- What moments of truth are likely to be most important to the client during the point of development phase?

POINT OF COMPLETION

Some IT organizations act as though the interaction with the customer ends when the problem is *"solved."* Actually, there is a third stage, rich with moments of truth, that often goes unnoticed by the service organization. This third phase is where we confirm that everything that we *think* we have done to resolve the client's issues *really has* resolved those issues … and we concisely "debrief" with the client to make absolutely certain that *the client feels* he or she has gotten the best from our team.

We should ask ourselves:

- Do we really have a set of procedures in place for the point of completion stage of the service relationship? Or do we simply "close the books" and assume that what we have done has worked wonders for our client?
- How do we know how clients feel about what we have done?
- If a client is still dissatisfied after we "close the books," what are that client's options?

- Who is responsible for the point of completion stage of the relationship?
- How are we gathering and analyzing information from clients at this stage that will help us determine what we are doing right ... and what could still use some improvement?

Last, and perhaps most importantly:

- What moments of truth are likely to be most important to the client during the point of completion phase?

THREE POINTS!

POINTS OF *CONTACT*

POINTS OF *DEVELOPMENT*

POINTS OF *COMPLETION* ...

AFFECT OUR CLIENT'S PERCEPTION OF OUR TEAM

THE FIRST THING CLIENTS WANT FROM US

CHAPTER EIGHT

**Taking a look at ourselves, we must ask:
What do clients really want from us?
And ... can we actually provide it?**

NOW THAT WE HAVE A better understanding of the potential impact of
our own communications with clients, let's look a little deeper.

What is the single most important expectation a client is likely to
have of us during one of those critical *moments of truth?* Does any one
expectation stand out?

As it happens, there is something that clients not only expect
from us, but feel they have a *right* to receive from us. After researching
clients' needs and expectations over a period of decades; after having
traveled not only in the United States, but also Canada, Europe, and
Africa; after evaluating thousands of IT organizations, I have been able
to put together a fully comprehensive expectation list that (I believe)
accurately describes exactly what IT clients typically expect from us.
And since we're trying to operate our IT operation a little bit less like
a department and a little bit more like a free-standing company, we're
going to want to become very familiar with the items on that list, and
particularly with the item that *heads* that list. Why? Because, as a stand-
alone operation, we need to work hard at meeting all of our company

operating expenses and (we hope) show a profit when the yearly books are closed. Like most businesses, we are going to have to compete for our clients' business. And that means meeting or exceeding client expectations.

Guess what? We can't meet our clients' expectations if we don't know what they are!

Let's begin with a few assumptions. They are as follows: we have already identified with our fictitious company a client base that will return or exceed the revenue we need to maintain our ongoing operational cost and show a *small* profit. The problem is, we cannot afford to lose *any* of our current, established clients. Why? Because we actually need to *grow* our business at a minimum rate of 10 percent per year—and it's much easier to expand our income from existing clients (without raising our fees) than it is to get that business from new clients! If we do that, there will be financial incentives for all of us here at Innovative Technological Systems. (Translation: Paid bonuses!) To help you meet this ambitious goal, you have secured the services of a strong consulting firm to advise you. And after spending a fair amount of money on this consulting firm's services, it turns out that they give you advice that is startlingly similar to the advice I've given you here. In order to hold on to your current clients and attract more business, you must be able to answer this question: **What *do* your clients really want from IT?**

The consulting firm you've engaged urges you not to take anything for granted, not to assume you know the answer to this question. For a small (okay, not so small) fee, the consultant agrees to do some work on your company's behalf, conduct an extensive, scientifically valid survey, and find out exactly what it is that is most important to your customers and prospective customers. After spending some more money and some more time, you have another meeting with the consultant, who hands you an envelope that contains the number one objective of your typical customer.

And, as it happens, that answer you've paid all that money for, and waited all that time for, happens to match up with the answer you're going to get from me when you turn the page.

Here is what the consultant's (expensive) survey said that your typical customer's top priority is:

R-E-S-P-E-C-T!

The number one requirement that always came to the top in all of the surveys I've ever done on this vitally important issue was the need for IT people to truly *respect* their clients.

Over and over again, our most valued clients and potential clients say that what they want most from us is to be *respected.*

Well, that seems easy enough. We would probably all agree about the importance of respecting our clients. Many of us may very strongly feel that we *do* already respect our clients. But do we? Let us take a closer look and see what clients really mean when they say, "I want respect."

When clients tell us that they want to be respected, what they mean is that they want to be heard *and listened to* ... attentively ... without ever being interrupted or disrespected!

Hmm ... if that's how our customers are spelling "R-E-S-P-E-C-T" ... are we really delivering it to them? The consultant asked that question, too, and it turns out that, overwhelmingly, our clients and prospective clients *do not think* we are doing a good job of listening to them. That feels a little odd to us, because we actually thought we were doing a pretty good job in that area!

The consultant's study clearly shows that clients want and actually demand that their views be truly listened to before we quickly dismiss them. Not only that—they want to be active partners with IT throughout the entire interactive process. They want to be greeted with respect at the point of contact, included as equal partners at the point of development, and kept informed, with no surprises, at every step of the process. Finally, they want us to confirm, with complete respect, that the job has been finished ... to *their* satisfaction ... to the point of completion.

Do we really think we're doing that for our clients, right now? Are we delivering on their number one expectation and making sure every *moment of truth* they experience with us sends the message, "We respect you"?

If not ... we've got some work to do.

I cannot stress enough, that respect—or, more likely, the lack of it—is most likely the number *one* negative *moment of truth* that creates powerful negative impressions about us, or IT as a whole, right from the start!

I'll be sharing a lot of strategies with you later on in the book about how to ensure that you send the right R-E-S-P-E-C-T message to your clients. Right now, though, here is a simple approach that will help you show them respect each and every time you interact with them.

Before any client meeting, *we* must take the initiative to learn a little something about *the* client's area of responsibility. *We* must learn a little about what issues this person has to deal with on a daily basis. *We* must learn a little about their successes and setbacks in life. *We* must go out of our way to learn a little about how the person, or people, we will be dealing with actually operate, and we must do so early in the relationship. *This is the simplest and most effective way to show R-E-S-P-E-C-T to our clients* ... and that's the number one thing they want from us.

R.E.S.P.E.C.T.

IT'S THE **NUMBER ONE THING** OUR CLIENTS WANT FROM US!

SO -- WE WILL GIVE IT TO THEM!

WHAT ELSE DO CUSTOMERS WANT?

CHAPTER NINE

THANKS TO ALL THE GREAT work turned in by our high-powered, high-priced consulting firm, we now know the number one deliverable we're responsible for: *respect*. Now the time has come to ask, "What else is on that list of deliverables we've commissioned?"

The next item that customers want from us, according to the survey the consulting team conducted on our behalf, is …

RESPONSIBILITY

Customers want us to act responsibly on their requests and needs; they don't want things to drop through the cracks. This is huge. In fact, there was a very close finish for the top spot: **respect** barely nudged out **being responsible** as the most critical element our customers demand from us.

Since we're running a free-standing, customer-centered business, we want to know exactly what responsibility entails. Fortunately, our consultant has conducted an extensive survey that illuminates all the key points. Here's what they found: *It means responding to customers.*

If we in IT can't deliver on our customers' requirements—can't adhere to the schedule and the standards that customers expect or have negotiated with us ahead of time—then the customer expects to be informed of the problem ahead of time. That's the *responsible* thing to do. Customers *do not* want to hear, "Sorry, I just had a terrible vendor problem, and I could not get back to you for a couple of days." That's not *responsible!*

Clients simply want to be kept in the loop. That's the essence of this requirement. At a bare minimum, to follow through on the expectations of responsibility our customers have, we should circulate and live by this simple policy: **Thou shall not let thy clients hang there uninformed. Period.**

Being responsible does *not* mean that we must have all the answers during the first discussion with the customer. It just means that we will commit to providing our best understanding of the very best way to handle the client's issues, as we understand them, in as honest and professional a manner as possible. As part of that effort, we may set up a time when we will touch base with the client again at a specific point in time to bring him or her up to date—and, in fact, it's highly recommended that we do this. Of course, if we make that commitment, we must live up to it!

Think about this part of the responsibility equation for a moment. When was the last time that you actually enjoyed waiting for a plumber to confirm the time and date he would show up? When was the last time you enjoyed staying at home all day so you could be there when the cable guy materialized to install the system? When was the last time you enjoyed having the appliance company suddenly, and without notice, shift the time when it was delivering your refrigerator? Most of us have some real horror stories about these kinds of events, horror stories we've made a point of sharing with other people: "They could have at least called and informed me of the change instead of just leaving me sitting there!"

Or, to bring it a little closer to home … how did you feel the last time you showed up for a scheduled meeting and found yourself all alone? After ten minutes or so, you did some digging, only to find out that the meeting had been changed. You had gotten no notice.

What was your reaction?

"Well, no one told *me!*"

Is that really where we want to leave our clients?

If we do a better job of being responsible, a better job of keeping people in the loop, a better job of keeping our commitments, we can greatly minimize our negative moments of truth … starting today!

CERTAINTY

In addition to the top-two finishers on the list, which must be our critical priorities as we build the plan for our client-centered business, that fancy consulting firm also gave us some other items to look at. It turns out that our customers are also very concerned about …

Certainty. Customers want to be *certain* that they are dealing with people who are knowledgeable, sincere, courteous, and professional. They also want to be *certain* that the people they've connected with will fulfill some other basic requirements. I found 4 words in the dictionary that I believe strongly have a direct impact on our ability to be service oriented when dealing with our clients. I call those basic requirements "The Four Cs." I utilized these "Four Cs" back in the 1980s, when I owned and operated my previous consulting firm, Ouellette and Associates of Manchester, New Hampshire and included them in my previous book, IS at Your Service. (Ouellette, L. Paul (1993)…IS at Your Service. Dubuque, IA: Kendall/Hunt Pub. Co. They are just as relevant today as they were back then. Pay close attention to these "Four Cs" with every client interaction!

Clients want to be certain that you show …

Confidence (without arrogance)
Competence (without overdoing it)
Commitment (without hesitation)
Consistency (without exception)

They're always looking for *certainty,* for *evidence,* of these "Four Cs"— in *each and every interaction* with our organization. Let's look a little more closely at each of the "Four C" elements in depth —so we're in a better position to give our customers the certainty they're looking for.

1. Confidence. How does our IT organization display confidence in general? Is it the right kind of confidence, the kind that supports people at all knowledge levels? Or do we project a little arrogance with anyone who simply does not understand us at the level that we feel that they should? Do we tend to show impatience when we're asked the same question several times? Do we "take charge" of the conversation when we feel that things are not moving as fast as we think that they should? Are we sensitive to the fact that our clients often do not understand our jargon? Do we take time to explain terms an outsider might not understand?

2. Competence. Do we demonstrate competence without being a little puffy? Do we take the time to demonstrate to our clients that they really are in good hands with us? Do we work at showing our clients that we can deliver exactly what we promise? Do we show that we do have the "total picture" in mind as we move forward on our commitments?

3. Commitment. Do we show a real commitment towards our clients? Do we make it known that we are there to serve them? If things do not turn out right initially, do we put the emphasis on making it right, or do we find reasons and analyses for the problem that prove it really wasn't our fault? Have we ever implied, or said outright, that the fault lay with the client?

4. Consistency. Do we show a consistency in our behavior toward our clients that allows them to accurately predict our approach to solving difficulties? Do we show respect on a consistent basis? Are we consistently on time for our client meetings? Is our professional behavior consistent? Does the client know what to expect from us? Do we do the right things—not just some of the time, but virtually every time?

Executing these "Four Cs" is a big part of our overall initiative for achieving a successful, professional, service-oriented IT organization. If we find ways to give clients certainty that we're "on track" in all four of these areas, we'll be going in the direction we need to go.

ACCESSIBILITY AND DEPENDABILITY

Accessibility and Dependability. Rounding out our list of survey items are two closely interrelated deliverables that I'd like you to think of as one: accessibility and dependability. This is the customer experience as it is actually delivered. And the expectations in these areas overlap.

Clients not only expect us to be accessible to them, but they expect all the access we offer them to reflect complete professionalism. This can be as simple as a prompt return phone call when someone can't reach us directly—but we have to remember that the *quality* of the access we give our clients is just as important as the speed. For instance, our clients expect to walk into our work area and notice that it is organized, clean, and neat. Similarly, clients expect e-mails and memos they get from us to be well written, courteous, and free of errors. They expect our functional equipment to be operational and reliable. They expect access, not just to anyone, but to people who speak professionally, who take their attire seriously, and who are professionally presentable.

Simultaneously, our clients expect to be working with people who are reliable—not in theory, but in the real world of actions and follow-through. That means they want to work with people who do not make promises that they cannot deliver, and who do exactly what they say they will do, each and every time!

The two halves work together as part of the same equation: perception. If either half is missing or inadequate ... the equation breaks down.

CLIENT EXPECTATIONS: HOW ARE WE DOING?

So—with those items on the list, how would you say our little company, Innovative Technological Systems, is doing thus far? As in, today? As in, right now?

Where Are We Now?

It's time for a reality check. Look at the whole list again, and then give me your honest assessment of where you think your IT operation is right now, on a scale of one to ten, when it comes to ...

Delivering **respect** to our clients: 1 2 3 4 5 6 7 8 9 10

Acting **responsibly** with our clients: 1 2 3 4 5 6 7 8 9 10

Providing **certainty** to our clients: 1 2 3 4 5 6 7 8 9 10

Being **accessible** and **dependable** for our clients 1 2 3 4 5 6 7 8 9 10

HIT THE BULLS-EYE!

DELIVER *RESPECT*

ACT *RESPONSIBLY*

PROVIDE *CERTAINTY*

BE *ACCESSIBLE AND DEPENDABLE*

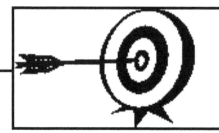

DIRT FOOT—AND LISTEN

CHAPTER TEN

No, THAT'S NOT A TYPO —the two big ideas we'll be covering in this chapter are **dirt foot** and **listen.** I realize they may sound a little funny when you read them together or say them out loud. As you'll soon see, though, these two ideas support everything that we've covered thus far … and they'll be a huge asset as we try to deliver on those items that we now know for sure our clients and prospective clients expect from us.

We'll start with the strangest-sounding of the two: **dirt foot.** Or, to be more accurate …

DIRTFT

This odd-looking acronym stands for Doing It Right the First Time! I don't know who coined the phrase but it has been around for a long time and I find it a good fit in a service environment. I pronounce it as *dirt foot* because I think it's a little easier to remember that way.

Now, I realize that you could never really expect to meet all of your IT requirements "right the first time" in each and every instance. You have lots and lots of variables to deal with, and not all of those variables lie within your control. So doing everything right the first time is clearly an unrealistic expectation.

So what are we really talking about when we talk about *doing it right the first time?*

What this means is that each and every time that any member of our IT organization interacts with any client, we are going to accept as a core principle that there is *no* excuse for that *interaction* not to be done right each and every time! That doesn't mean we complete everything on the list necessary to solve that problem perfectly and within the next thirty seconds … but it does mean that we take full responsibility for the interaction.

Take any of the components of the client relationship building that we have discussed so far and see how you can in fact make each client interaction successful … by Doing It Right The First Time. Yes, we can do *that* much right each and every time—simply by showing the other person that we're willing to make an effort!

This means doing things like showing courtesy, respect, patience, and other elements of compassionate human interaction. Can you imagine what we'd be able to deliver during our moment of truth interactions with clients if all we did was use what we've already learned here … to Do It Right The First Time?

LISTEN

Now it's time to look at the single most important component when it comes to building relationships that deliver the respect our clients demand, and everything else on that list of expectations they have for us. This critically important item is LISTENing, and it's the culmination of literally decades of work on my part in the area of client satisfaction in the realm of IT.

LISTEN is an acronym, too, of course. The acronym itself reminds us that we must give clients and prospective clients our complete and undivided attention when we talk to them. If we get only that much out of this chapter, we've already addressed a huge part of the *respect* equation! Beyond that all-important reminder, of course, there's more—a lot more—that LISTEN has to teach our IT organization.

I guarantee you that if you learn, practice, and implement what follows, you *will* gain complete respect not only from your clients, but from your significant other, your close relatives, your colleagues,

friends, and acquaintances, and anyone else you implement the LISTEN model with!

After learning what I'm about to share with you, one of my students came in to my program the next day and asked to address the other participants in the class. I agreed. He proceeded to tell us that the previous evening, while waiting for dinner to be served he followed his normal routine and was sitting at his workbench, doing some woodworking.

Just then, his six-year old son, who had been playing outside, came in and asked if he could share something important that he had accomplished that day. My student (the father) stopped and reflected about the new listening tool that he had learned in my class that day. He turned off his equipment and turned to his young son, looked him in the eye, and said "Sure, please tell me about what you accomplished today."

Upon seeing this, his son ran to his mother, saying, "Mommy, Mommy, Daddy's *listening* to me!"

With tears in his eyes, he told the class, "My son had ridden his bike without training wheels for the first time that day—and that was paramount to him. I realized that for my son's entire life, I had not *really* been listening to him. I have now changed that. Thank you, Paul."

Let me ask you a question. Out of all the people you have known in your life, how many would you classify as *truly great* listeners, people who consistently and unhesitatingly gave you their full and undivided attention? Think of the actual names—or, even better, write them down on a separate sheet—before you continue with this chapter.

How many names did you come up with? One? Two? More? During my training sessions over the past twenty-plus years, my experience is that most people cannot think of a single person who actually fits the description I've just laid out, and that those who can—fifteen to twenty percent of my students—can only think of one or perhaps two people. I ask participants who actually do think of names to answer this question: "How do you feel about those people?" The answer I get is always some variation on the following: they like the person, they respect the person because they know the person respects them, they admire the person—

or a combination of two or all three of these elements. As human beings, we all love it when people truly listen to us!

The moral: If our goal is to show respect ... we must first LISTEN!

Here's how the LISTEN model breaks down:

L stands for *listening*.
I stands for *interest*.
S stands for *sincerity*.
T stands for *trust*.
E stands for *empathy*.
N stands for *nonverbal* and *never*.

L Stands for "Listen"

The first letter in the LISTEN model stands for listening itself.

There are two components to consider here: The first one is the sound that we may hear from the speaker, and the second is the sound of our surroundings.

We have to clear our minds of any distracting or disturbing sounds that are competing with the message ... so we can hear the sounds the person is making! Otherwise, we won't be doing a very good job of listening.

We have to make an effort to reduce or eliminate "background noise" or "white noise," perhaps by moving elsewhere, or by turning off the source of the sound (though this is not always possible), or by closing the window or the door, or, if all else fails, by making a conscious effort to blank the competing sound out of our minds. This is not as impossible as it may seem at first; with conscious effort and a little practice, we can learn to keep external "white noise" from competing with the sound of the speaker's voice.

Even after we remove or minimize the influence of external "white noise," we are not done. We must also recognize and manage our own *internal* "white noise," which can be even more distracting! You know what internal "white noise" is: thinking about what we're going to say next, or what we're going to have for lunch, or who this person reminds us of, or maybe even what we would *really* like to tell this person to do. Internal "white noise" is the cumulative sound in our own heads that negatively affects our own ability to listen to the speaker. Sometimes, we feel pretty sure that we can "multi-task" and do two or more things at any one time in our mind—like, for instance, listen to a customer at the same time we solve some other problem. Why do we believe we can do this? Well, it's not because we're bad at dealing with customers. It's because, for most of us, our brain tends to move at a speed that's roughly five times greater than the customer's ability to speak. That gives us some spare moments, and, because we want to make the best possible use of our time, we try to fill this "unused capacity" with something else.

That's what we tell ourselves … *but there's a problem.* This little internal justification is one of the big reasons we are not good listeners. In fact, it might just be the biggest reason!

We must make every possible effort to clear our minds and concentrate fully on what the speaker is saying—that and only that. The human mind is an amazing thing. After only a few days of practice of giving someone your complete, undivided attention, you will be able to pave the way for a much more effective listening routine. With just a little effort, most of us can devote up to two minutes of complete, focused concentration on what the speaker is saying before we become saturated. The trick then is to use your "spare capacity" to quickly summarize the key points that the speaker *has just said* … and then immediately return to the listening process.

"People ought to listen more slowly."
—Jean Sparks Ducey

If you've ever wondered whether someone was really listening to you … then gasped in astonishment as that person repeated back all the major points you just shared … you now know what was happening. When that person felt the urge to "multi-task" by thinking about some other topic, his or her attention went instead to the job of summarizing, mentally, everything you'd been talking about! You can learn to do that, too … with just a little sustained practice.

I Stands for "Interest"

Most of us, after thinking that we have heard everything we need to know about what the speaker is saying, just can't wait to get in there and deliver our answer, our opinion, or our valuable expertise. So what do we do? We cut the person off. Whenever we do that, we send this message: "I have absolutely no interest in what you're saying."

When we "jump in," we show that we have lost patience—and interest—with both the person and the topic as he or she is describing it. We want to wrap this up, here and now.

Wrong answer! The person we're talking to will *always* pick up on our urgency to close out this conversation, and this desire to "wrap it up" will *always* be more significant to our conversational partner than any solution we might suggest. End result: *no respect!* We aren't *showing* any respect to the customer we're talking to, and as a result, we certainly aren't going to get any in return.

We may think that we have done a great job of listening when we try to close out the conversation early, but we really haven't. By abandoning real *interest* in the person and the topic, what we are really guaranteeing for ourselves is some variation on the following exchange:

Customer: What about X?

Us: X? Did you say anything about X?

Customer: Yes. Don't you remember?

Us: Really? Are you sure?

Of course, the customer really did say something about X—we just didn't hear it because we were too busy figuring out how to solve the problem based on the information we already had at our disposal. *We lost interest, and as a result, we missed an important piece of the puzzle.*

S Stands for "Sincerity"

If we want to show full respect to those we listen to, and win respect in return, we must spend time on the next step in the listening process: sincerity. The speaker needs to know that we are making a sincere effort to listen—that we are doing our level best to get all the information we need in order to solve the problem.

That means we never place any kind of judgment on any part of what the speaker is saying; instead, we accept it at full face value. *What the customer is sharing with us* is *the customer's experience, and that experience matters.*

We must demonstrate to this speaker that he/she has a full right to those statements, as they stand. When we do this, the customer begins to feel a freedom to express views openly, because we are not passing judgment on, or rebutting, what's being said. Once the customer knows that we are truly sincere in our efforts to get all the information … something amazing happens.

T Stands for "Trust"

Whenever we really listen, whenever we show genuine interest, whenever we make it clear that we are willing to make a sincere effort to gather all the information, including the client's subjective assessment of what happened, without judging or rebutting, then **trust** can take root.

The speaker needs to know that he or she can trust us. This is a "reality check" step. If we have followed the listening model so far, the speaker will automatically feel that trust and will respond in kind. On the other hand, if we *don't* feel trust growing in the conversation, the odds are very good that *we have not been following the LISTEN model at all!* We have to go back to the beginning and start over!

E Stands for "Empathy"

Empathy means understanding and respecting the feelings the person has been experiencing, and even sharing in those feelings a little bit. If we really hope to gain respect from those that we listen to, we must uncover all of the feelings that the speaker is trying to convey to us and make it crystal clear that we "get" this part of the picture.

If we have all the facts right, but have no sense of how the person *feels* about what's been going on, we have not done our job. There's no way around it: We need to engage with the other person's interest, passion, energy, beliefs, happiness, sorrow, agreements, disagreements, feelings of support, feelings of lack of support, and so on. If the emotional piece is missing, we have not "gotten" the true message that the speaker is trying to convey to us.

Often, empathy is as simple as saying, and meaning, something like this: "That must have been very frustrating."

Identifying with your speaker at this emotional level is one of the very best ways to connect—and once you connect in this way, you receive respect from the speaker. Your agreement or disagreement about the "facts of the case" at this point is really of no importance. Your job at this stage is to simply *show that you truly understand, and identify with, the person's feelings.*

Don't skip the empathy step!

"N" Stands for "Non-Verbal" … and "Never"

This last letter of the LISTEN model carries two reminders.

The first is that we must always check ourselves for **non-verbal** messages that undercut the LISTENing process and show a lack of respect. For instance, if we're trying to show empathy for the other person's emotional response, but our face is saying "Who cares?"—there's a big problem. We must never forget that although the speaker is engaged in his or her talk, the customer is also looking at us and constantly monitoring our reaction. The person we're talking to *will see*—and process—every non-verbal cue we send. This is because *most* human communication is delivered by means of non-verbal cues! Maintain comfortable eye contact with the speaker at all times, and make an effort not to send any negative visual message with your face, with your posture, or by staring at something else. (There's a lot more to say about body language and eye contact; I'll address these issues in much more depth in the next chapter.)

The second big message here is that we must **never** attempt to respond to the speaker until we are absolutely certain that the *speaker* feels he or she has completed his or her message. Allow for a little "empty space" before you launch your rebuttal.

Are You LISTENing?

There you have it—the key to true LISTENing. Practice this LISTEN sequence; try it with your best friend or significant other, and watch what happens. You *will* get better and better at the process, you *will* gain great respect, and you *will* feel great about yourself in the process. That's because listening well is a life skill, and the better you get at it, the better your life relationships will be.

As you will see in the next chapter, mastering the LISTEN model is really the work of a lifetime. That's because we human beings are constantly "saying" things … when we think we're not saying anything at all.

REMEMBER...

DO IT *RIGHT THE FIRST TIME* ...

AND, WHEN IN DOUBT, *LISTEN!*

WHAT WE MAY BE SAYING WHEN WE'RE "NOT SAYING ANYTHING"

CHAPTER ELEVEN

ONE SUNDAY, WHILE OUT WITH friends, I learned, yet again, that I simply couldn't trust my own face.

I was sitting there, carefully listening to a lecture when Linda, who was sitting next to me, turned to me and said, "Paul, is the lecture boring you?"

"Absolutely not." I replied. "I'm captivated."

At which point, Linda said, softly but firmly: "Well, *tell your face!*"

It had happened again, all too easily, and without my even realizing it—just as it had happened to so many others before me, and just as it was likely to happen to "autopilot" service people for years to come. My face was telling a very different story from the story I thought it was telling. It happens to all of us: If we don't make a conscious choice to send a positive message, we can send a negative one without even thinking that's what we're doing.

Here's the moral of the story: **autopilot is too expensive.**

As professional service providers, we have an obligation to bear in mind the possibility that we are "saying something" without meaning to "say something." And ... we must constantly be on guard against sending one particular, extremely powerful nonverbal message: the DILLIGAD look.

You remember: the "Do I Look Like I Give a Damn?" look. I touched on this facial expression a little earlier in this book, but it can always use a little further discussion.

This DILLIGAD look is a real turn-off for our customers. In fact, it's *such* a turn-off that we have a professional obligation never, ever to use it while we're on the job … regardless of what we say or don't say with our voices! *The DILLIGAD expression can and will poison a moment of truth, even if we are doing everything else right!*

We can confirm this simple, reliable principle for ourselves very easily. Each of us has encountered the store clerk, the waiter, the salesperson, the ticket-taker, the friend of a friend—*someone* who gave us the DILLIGAD look and changed our emotional state without even saying a word.

What did we want to do when we saw that expression?

I'll tell you what we wanted to do. We wanted to lash back, right then and there! But in all likelihood, being the nice people that we are, we didn't do that. Instead, we simply walked away and started thinking about how to avoid ever having to come in contact with that person again—unless we absolutely, positively, had no choice.

Sometimes, I think it might be better for us, as professional service providers, if our customers *did* lash back every time we (inadvertently) flashed a DILLIGAD expression. At least that way, we'd know when we were doing it … and we'd also have some strong negative reinforcement that might help to prevent us from giving the expression again without realizing that that's what we were doing!

The DILLIGAD expression simply will not work to our benefit. We must be incredibly careful about which facial expressions we're circulating, because they say so much about us without us realizing it—while we actually think we're not saying anything at all!

OTHER POTENTIAL "WHAT WE'RE REALLY SAYING" PROBLEMS

Here are some other areas where we can "say things" without realizing that that's what we're doing.

Our attire says things to the client. Our attire needs to be appropriate for the occasion, neat, and well-coordinated. That goes for such items as your belt and shoes. Shoes should be clean and have no scuffs. There should be no shirt-tails hanging out, no wild colors, and no short hemlines. If your organization has a "casual day" or "dress-down day" once a week, make sure that the messages your casual, dressed-down attire are sending to clients are the messages you want to send.

Our grooming says things to the client. Hair must be neat, trimmed stylishly, and clean. Women whose makeup is applied unevenly or who use lipstick and blush shades that are too bright will run the risk of distracting from the message they're trying to send.

For men, our facial hair says things to the client. Beards and mustaches need to be trimmed and neat. This includes eye brows, the hair along the neckline, and hair around the ears.

Our jewelry says things to the client. If we wear jewelry, it should be professional-looking and appropriate to the occasion. Gaudy jewelry sends the wrong message entirely and will distract people from everything we're doing to support the relationship. Less is more!

Our accessories say things to the client. Our briefcase, for instance, should be stylish, neat, and businesslike. It should have no rips or broken handles; it should not be overstuffed with papers or other items. The same goes for handbags.

Even our glasses say things to the client. They must be stylish and neat, with no smudges on the glass itself.

WHAT WE SAY WITHOUT TALKING MATTERS!

Please take this seriously. As consumers, we really do read messages into facial expression, attire, grooming, and other (apparently) "silent" means of communication when we are on the receiving end of the service exchange. As service professionals, we must be sensitive to the kind of messages we are sending out. If those messages don't support our service philosophy, our customers will make *negative* judgment calls about us and our organization based on what they "hear" in our silent—and all too often unconscious—messaging. And their *negative* assessments will persist for years or even decades! Whether or not this seems fair, this is the way human communication works.

So much for the messages *we're* sending. In the next chapter, we'll look at the messages the *customer* is trying to send to us!

WATCH IT!

SOMETIMES, IT'S WHAT WE SAY WHEN WE'RE **NOT TALKING** THAT MAKES THE DIFFERENCE TO THE CLIENT!

ACTIVE LISTENING TAKES PRACTICE

CHAPTER TWELVE

YOU AND I HAVE NOW spent some significant time together looking at what, exactly, must happen for our customers to perceive our little "company" as being a truly service-oriented team. And by this point, it should come as no surprise to either of us to learn that one of the critical skills we are going to have to develop, strengthen, and use over time if we hope to support that stand-alone, customer-focused outfit of ours is *listening*.

Listening—really listening—makes the difference and will give us a significant competitive edge. But it takes practice. In a previous chapter, we learned about the LISTEN acronym. In this chapter, we will learn how to put active listening into practice.

The kind of listening I'm talking about is not a passive activity, at least not for the team members of our new "company." *Our* brand of listening must be active. In other words, we must look for ways to *encourage our clients to voice their feelings to us* about the service we are delivering to them.

Of course, not everyone in "tech support" does this. Our customer-focused organization, however, is going to specialize in it.

We have to make our listening active in this way. We have to go out of our way to figure out how our customers really feel about what we're doing for them—or not doing for them Why? Because ...

Good news travels slowly ... but *bad news* travels very *fast!*

To put it bluntly, active listening is a *strategy* that helps us to contain, and turn around, "bad news" experiences before they reach the outside world. If there's a problem, we want to hear about it, find out what it is, and resolve it ... before anyone else hears about it.

Now, I know that identifying problems that our customers may have with what we've said and done is not always the first thing we want to do. But believe me, once we start implementing the processes I'm about to share with you, once we start noticing the effect that our active listening efforts are going to have on our relationships with our customers and our team's bottom line, we will wonder how we ever got along without these tools. By using what follows, we will identify our clients' real needs and find out what they really need from us—not just from the technical system point of view, but also how they expect us to respond to them as one human being to another.

BEING OPEN TO COMPLAINTS

You read right. We must be open to customer complaints. In fact, our goal is not only to be open to listen to customers when they feel like complaining, but actually to *seek these complaints out before* they become deeper issues. To do this, we must create a complaint management system that's customized to our organization. This system must be part of our standard operating procedure, and everyone on the team must know how to administer it whenever we are successful in unearthing a customer complaint.

Perhaps you're wondering: Why do we have to bother creating such a system?

Here's the answer. There are countless organizations today that carefully measure the effects of poor, average, good, or great service on customer loyalty and referrals. The research on such companies shows that those who provided a high level of service saw an increase of five percent in their business—even though these organizations charged, on average, *nine percent more for their products and services* than their competitors. That result in itself is a very powerful case for the benefits of being service-oriented as an organization ... but there's something else of critical importance for us here. The same study also revealed that *only four to five percent* of the customer base would complain when they knew for certain that they were dissatisfied with the service they'd gotten.

Think of the implications here: *Most of your customers, approximately 95 percent, won't tell you when they're dissatisfied.* Many of these people will simply walk away and never return!

WATCH OUT!

We know for sure that **most of our unhappy customers who have problems with us never bother to complain … which means their complaints go unseen and unnoticed.** It's our job to do something about that!

Don't fool yourself into thinking that our organization has a "captive audience" that can't or won't walk away. Remember: *seventy percent* of customers who leave do so because of some kind of negative interaction with the seller!

Let's face it. We're in trouble if we don't go out of our way to *elicit* complaints and problems. The vast majority of our customers won't complain when we mess up a moment of truth, which means that the vast majority of our customers are likely to walk because of a problem we never found out about! Think about that.

We may fool ourselves into believing that we're immune to the syndrome of losing a customer to the competition, and so don't have any obligation to figure out what kinds of complaints customers aren't sharing with us. If we fall into this myth, we must rejoin the reality-based community … quickly! With the advent of ever-more-specialized, ever-easier-to-use software, it's only a matter of time before someone takes advantage of our complacency—and creates a better, simpler, more up-to-date offering that's customized more closely to the needs of our customer. The vendors of these products are working overtime to sell directly to your customer base … and they are looking for ways to treat customers *better than we are currently treating them.* Again, studies over time in corporate America show that approximately five percent of people will complain directly to the service provider about the quality of their experience—only five percent. That's not anywhere near enough, but at least we've got a chance of repairing the interaction and making the necessary adjustments with those five percent. More importantly, though, we need to notice that an astonishing forty-five percent of our current customer base is actively *spreading the bad news* by complaining to someone else, instead of us! That's not good for our image at all, and we have only a very slight chance of fixing the relationship and making the necessary adjustments with these folks. Why? Because we don't know what the problem is!

As if that weren't enough, an additional 50 percent of our customer base is not complaining at all, but would probably love an opportunity to try … or go someplace else when they see an opening. This is a *huge* competitive challenge for us … and we have to come up with a response to it.

Believe it: We must go out of our way to *generate* complaints. And to do that, we must build an effective complaint policy into our overall service strategy. I'll show you how to do that in the next chapter.

Reasons customers choose not to complain to us, even when they know they have a problem:

1. It is simply not worth the time or energy.
2. No one really cares or is willing to take corrective action.
3. There is no known way to express dissatisfaction.
4. Customers are afraid that their complaints may well have long-term negative effects on current or future projects.

LISTEN ACTIVELY!

YOU CAN **PREVENT FUTURE PROBLEMS** WITH ACTIVE LISTENING. REMEMBER: BAD NEWS TENDS TO **TRAVEL FAST** ... AND GOOD NEWS TENDS TO **TRAVEL SLOWLY!**

BUILDING OUR SERVICE STRATEGY

CHAPTER THIRTEEN

THERE ARE FIVE ESSENTIAL STEPS to building a powerful, customer-focused service strategy … and there are also a number of myths that can keep us from taking these steps in our organizations. In this chapter of the book, we'll look at the five critical steps and what will make it possible for us to carry them out. What's more, we'll see why the myths that are offered in opposition to creating and sustaining such a service strategy simply do not withstand scrutiny.

Let's begin with the five essential steps.

STEP ONE: WE MUST ACCEPT THAT "JOB ONE" IS TO PLACE CLIENTS AT THE CENTER.

Our strategic goal is and must always be to create a service-oriented organization. That means placing and keeping clients at the center of our every decision. This philosophy is non-negotiable; it has to be the core of our service strategy. We must place the customer at the center of all our efforts and make sure that customers not only know intellectually that they are at the center of our efforts, but also *feel*, in every interaction, that they themselves are the driving force of what we're doing and why we are doing it.

STEP TWO: WE MUST ARTICULATE A SERVICE PHILOSOPHY FROM THE TOP DOWN.

Absolutely everyone connected directly or indirectly to our team must articulate a clear service philosophy—one that is unique to our business and its customers, consistent, competent, and fully supported by top management. In other words, *everyone* in the organization must know that they will be rewarded, and never penalized, for putting the customer first and specifically for identifying and escalating complaints (see Step Four). This point is of such extraordinary importance that I hope I will be forgiven for emphasizing it throughout this chapter. Our team must get clear evidence for the belief that identifying and acting on complaints is a good thing to do, and they should get that evidence from the most senior people in our company. Our most important corporate leaders must express a powerful, service-centered operating philosophy on a *weekly or daily*—not quarterly or yearly—basis.

STEP THREE: WE MUST DEVELOP AND SUPPORT A STRONG, SERVICE-ORIENTED STAFF.

Our staff must be recruited, trained, rewarded, and retained based on their strong service focus. (Making sure everyone on our team reads this book is a good place to start.) We must create a strong customer-focused working culture, not just once in a while, but consistently over time. In creating and sustaining that culture, we must support each other with *comfort* when needed, with *clarification* when required, with *confrontation* when applicable, and with full-scale *celebration* when our clearly established service goals are finally achieved.

All four of these elements are essential, because being a daily service provider is, quite frankly, not an easy task. To keep the staff rightfully energized, we need all four of these responses to be part of our ongoing business plan and our daily management routine.

Ouellette Solutions offers a two-day training program on the topic of consistent service delivery. I have worked with over a thousand organizations over a period of more than twenty years who have participated with great success. To find out more, contact Ouellette Solutions Inc. at www. ouellettesolutions.com or <u>speakercoaching@yahoo.com</u>. or call 239-514-1478

STEP FOUR: WE MUST CREATE AND SUPPORT A COMPLAINT ESCALATION POLICY.

In keeping with the ideas we examined together in the previous chapter, we must develop a process for identifying customer complaints—so that they can be spotlighted and resolved. This, too, is non-negotiable. We must have a policy that clearly allows us and our clients to escalate a problem to a higher source for resolution if it is deemed that the resolution or approach at the current level is not satisfactory. Again, this policy must never, ever result in retribution or retaliation to any team member. To the contrary, some sort of reward structure should be in place for our ability to respond to complaints in a way that actually helps complete an ongoing project to a customer's complete satisfaction—or even prevents a major disaster from happening.

Sometimes, you may not have the resources or the political power you need to resolve a problem. You should have the *authority* to act quickly and decisively—and not be penalized for doing so. Ideally, you should be able to apply appropriate resources until the problem is resolved. This is the acid test for the organization as a whole. Top management must support IT's resolution of the complaint cycle! You should have the authority to take intelligent action until the problem is resolved. When people on the "front line" don't have the authority or the resources to solve a problem … that's when bad things happen.

STEP FIVE: WE MUST HAVE SERVICE MONITORS IN PLACE AS PART OF OUR INTERNAL SERVICE MANAGEMENT RESPONSIBILITY.

This means we must develop and install people I call "service monitors" on the job. These can be either new hires or current team members who are doing "double duty." Whatever their profile, these folks must have the clear, unambiguous, and personal responsibility to check into how well specific members of the team are handling the job of identifying and resolving complaints. Those who are acting as service monitors should measure things such as total complaints, the specific actions undertaken, and the customer's satisfaction level after complaints have been addressed. Then they need to evaluate our actions in each of these areas *considering the goals we have set for ourselves as acceptable service levels.*

PULLING ALL THIS OFF MEANS BECOMING MUCH MORE FLEXIBLE!

Those are the five critical steps we're going to want to carry out as a team. Accomplishing them will give us a competitive edge, because the five steps are *not* what most IT teams focus on.

Most IT teams get into a rut—they follow a routine and deliver about the same (unacceptable) level of perceived service all the time. In order to turn these five new ideas into realities on the ground, we're going to have to operate in a more flexible manner.

In fact, to hit our income and customer retention goals for Interactive Technological Systems and to match the achievements of the IT teams I've worked with as a consultant, I believe we are going to have to offer not just one level of service, but three. I call these three levels of service *data-driven, analytical,* and *decision-driven.* Let's look closely now at what goes into each of those three levels of service ... and why it's so very important to understand the differences between them.

THE "DATA-DRIVEN" SERVICE LEVEL

At this level, our focus is on the number of transactions produced. Our goals are those of speed and efficiency. Our aim is going to be to reduce the total number of steps required to achieve the best possible outcome for our clients. The emphasis here is on activities like managing volume during a period when there is an extremely high volume of inquiries.

THE "ANALYTICAL" SERVICE LEVEL

When we are operating at the analytical service level, our focus is based on the idea of delivering *convenience of usage during a single exchange with the customer.*

At this service level, our goal is to meet all the customer's specifications, as the customer defines them. Delivering quality, as perceived by the customer, is the key concern. We want to make sure that we are delivering, not just a successful call or transaction, but an overall service *experience* that our customer will remember positively. Even with this positive experience, however, the relationship remains basically transactional.

THE "DECISION-DRIVEN" SERVICE LEVEL

This is where we are focusing on *exceeding* expectations—not just meeting them—and on building the relationship over time. We are continually looking for ways to add value. Our goal is not just to deliver a single great transaction, but to be absolutely sure that the client is 100 percent satisfied with the outcome.

Different organizations have different service philosophies, and each of the above approaches has its advantages for certain types of operation. **My strong recommendation is that every client interaction *begin* at the decision-driven level, regardless of the client's position or status, until we have a system in place that is capable of supporting analytical, data-driven, or decision-driven systems successfully.**

We must know how to deliver all three "flavors" of service ... and we must learn, over time, which "flavor" makes the most strategic sense for both us *and the client* at any given moment. Identifying the right blend of Data-driven, Analytical-driven and Decision-driven is a big part of creating and implementing a Unique Service Philosophy (USP) that is responsive to our clients and "glove-fitting" for our organization.

ADVANTAGES OF A UNIQUE SERVICE PHILOSOPHY

1. It establishes and monitors the value our clients receive.
2. It informs our clients of exactly what they can expect to receive from us.
3. It keeps expectations in line within established, realistic guidelines.
4. It gives clients a voice and a process by which to respond to us.
5. It helps us to unify our service team.
6. It gives us a better idea of the expenditures needed on an ongoing basis to meet established service requirements.

... and perhaps most important of all ...

7. It improves our end result and generates a new level of respect from our clients ... and the realistic potential for increased revenue from the client base.

Developing this philosophy is so important to our organization that we should go the extra mile as we put it together. What do I mean by "go the extra mile"? As we develop this glove-fitting way of synchronizing our customers, our resources, and our processes, we should involve not only IT management, but also personnel from our client community and staff members from the various components of the IT organization. We must build these voices in right from the beginning, and we must let all of them have a say in our work as it goes forward. With the help of this diverse group, we are going to take on some very serious questions. Among them:

☐ What do our customers consider to be an acceptable level of service in X area?
☐ How close are we to meeting their expectations right now?
☐ What kind of commitment, specifically, do our customers expect from us in X area?
☐ Are we, as an organization, ready for those commitments?

☐ Will we be able to commit the funding that will be required for training our personnel to meet those commitments?

☐ Are we prepared for the concerns that may be raised by our staff?

☐ Do we have the appropriate staff levels to meet the objectives we have established for ourselves?

☐ How do we establish this new process within our client community? What will the first phase look like? What will the second phase look like? What will the third phase look like?

☐ How will this new strategy impact our established mission statement?

☐ Will the mission statement need a rewrite?

TO ANSWER THOSE QUESTIONS ...

We will probably have to do a sizeable service client assessment survey as a first step. An example of this kind of form follows.

ASSESSING OUR IT SERVICE LEVEL

Name _____

What is your impression of your organization's client service effectiveness? Check the number you find appropriate.

	Never	Not Often	Occasionally	Frequently	Always
1. We respond quickly to our clients.	0	1	2	3	4
2. We actively listen and ask questions that make our clients' concerns clear.	0	1	2	3	4
3. We deliver projects with few errors.	0	1	2	3	4
4. We deliver quality goods and services the first time.	0	1	2	3	4
5. We spend the time required with our clients to completely understand their situation.	0	1	2	3	4
6. We quickly respond to our clients' questions.	0	1	2	3	4
7. We explain projects in non-technical jargon.	0	1	2	3	4
8. We look professional and conduct ourselves professionally.	0	1	2	3	4
9. We deliver on our promises.	0	1	2	3	4
10. We enjoy providing our clients with individual attention.	0	1	2	3	4
11. We respond proactively to difficult situations.	0	1	2	3	4
12. We explain our processes to our clients so they understand and learn.	0	1	2	3	4
13. Our work areas are attractive and tidy.	0	1	2	3	4

ASSESSING OUR IT SERVICE LEVEL

Name: _____

	Never	Not Often	Occasionally	Frequently	Always
14. We understand our clients' businesses and their individual technology needs.	0	1	2	3	4
15. We plan carefully when and how to say no to a project or proposal.	0	1	2	3	4
16. We are eager to help our clients in any way we can.	0	1	2	3	4
17. We ask effective questions and give our clients confident reasons why we are asking them.	0	1	2	3	4
18. We deliver products without problems.	0	1	2	3	4
19. We finish projects on time or ahead of schedule.	0	1	2	3	4
20. We guarantee our product/service.	0	1	2	3	4

ASSESSING OUR IT SERVICE LEVEL

Name:

Questionnaire Assessment Results

Insert the number selected for each question 1-20 below and total each column.

Openness	Confidence	History of Success	Dependability	Emotionally Responsive
1. _____	2. _____	3. _____	4. _____	5. _____
6. _____	7. _____	8. _____	9. _____	10. _____
11. _____	12. _____	13. _____	14. _____	15. _____
16. _____	17. _____	18. _____	19. _____	20. _____

Totals _____ _____   _____   _____   _____

Chart the above totals below for a visual graph of service strengths/weaknesses:

Openness	Confidence	History of Success	Dependability Responsive	Emotionally
20	20	20	20	20
—	—	—	—	—
—	—	—	—	—
—	—	—	—	—
15	15	15	15	15
—	—	—	—	—
—	—	—	—	—
—	—	—	—	—
10	10	10	10	10
—	—	—	—	—
—	—	—	—	—
—	—	—	—	—
5	5	5	5	5
—	—	—	—	—
—	—	—	—	—
—	—	—	—	—

CLIENT ASSESSMENT FEEDBACK

From the Client's Point of View

Name:

Check the number that is your impression of the services you currently receive from IT.

	Never	Often	Occasionally	Frequently	Always
1. IT responds quickly to me.	0	1	2	3	4
2. IT actively listens and asks questions that make my concerns clear.	0	1	2	3	4
3. IT delivers projects with few errors.	0	1	2	3	4
4. IT delivers quality goods and services the first time.	0	1	2	3	4
5. IT spends enough time with me to completely understand my situation.	0	1	2	3	4
6. IT quickly responds to my questions.	0	1	2	3	4
7. IT explains projects in non-technical jargon.	0	1	2	3	4
8. IT is proficient in meeting client business needs with effective solutions.	0	1	2	3	4

CLIENT ASSESSMENT FEEDBACK

From the Client's Point of View

Name:

	Never	Often	Occasionally	Frequently	Always
9. IT delivers on their promises.	0	1	2	3	4
10. IT enjoys providing me with the attention I need.	0	1	2	3	4
11. IT responds proactively to difficult situations.	0	1	2	3	4
12. IT explains their processes to me so I can understand.	0	1	2	3	4
13. IT considers alternative solutions and looks for innovative ideas.	0	1	2	3	4
14. IT understands my business concerns and my individual technology needs.	0	1	2	3	4
15. IT knows when and professionally how to say no to a project or proposal.	0	1	2	3	4
16. IT is eager to help me in whatever way they can.	0	1	2	3	4

LIENT ASSESSMENT FEEDBACK

From the Client's Point of View

Name:

	Never	Often	Occasionally	Frequently	Always
17. IT asks effective questions and gives me confident reasons why they are asking them.	0	1	2	3	4
18. IT delivers products that are both cost-effective and free of errors.	0	1	2	3	4
19. IT finishes projects on time or ahead of schedule.	0	1	2	3	4
20. IT exceeds acceptable levels of customer satisfaction.	0	1	2	3	4

CLIENT ASSESSMENT FEEDBACK

What They Really Think

Name: _____

Insert the questionnaire number that you selected for questions 1-20 below from each of three respondents. (You can include up to three individual client responses.)

Openness	Confidence	History of Success	Responsibility dependability	Emotionally
1. __ __ __	2. __ __ __	3. __ __ __	4. __ __ __	5. __ __ __
6. __ __ __	7. __ __ __	8. __ __ __	9. __ __ __	10. __ __ __
11. __ __ __	12. __ __ __	13. __ __ __	14. __ __ __	15. __ __ __
16. __ __ __	17. __ __ __	18. __ __ __	19. __ __ __	20. __ __ __

Totals _____ _____   _____   _____   _____

Averages _____ _____   _____   _____   _____

Openness	Confidence	History of Success	Dependability Responsive	Emotionally
20	20	20	20	20
__ __ __	__ __ __	__ __ __	__ __ __	__ __ __
__ __ __	__ __ __	__ __ __	__ __ __	__ __ __
__ __ __	__ __ __	__ __ __	__ __ __	__ __ __
15	15	15	15	15
__ __ __	__ __ __	__ __ __	__ __ __	__ __ __
__ __ __	__ __ __	__ __ __	__ __ __	__ __ __
__ __ __	__ __ __	__ __ __	__ __ __	__ __ __
10	10	10	10	10
__ __ __	__ __ __	__ __ __	__ __ __	__ __ __
__ __ __	__ __ __	__ __ __	__ __ __	__ __ __
__ __ __	__ __ __	__ __ __	__ __ __	__ __ __
5	5	5	5	5
__ __ __	__ __ __	__ __ __	__ __ __	__ __ __
__ __ __	__ __ __	__ __ __	__ __ __	__ __ __
__ __ __	__ __ __	__ __ __	__ __ __	__ __ __

Client responses ___ ___ ___

CLIENT FEEDBACK ASSESSMENT: A TOTAL LOOK

Rating Ourselves for Client Empathy and Responsiveness

How would you, as an IT professional, rate your current IT organization as a total unit?

HIGH **(We empathize with our clients.)**

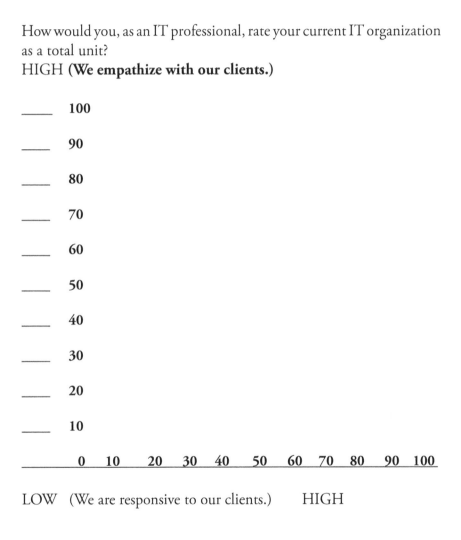

_____	100
_____	90
_____	80
_____	70
_____	60
_____	50
_____	40
_____	30
_____	20
_____	10

0 10 20 30 40 50 60 70 80 90 100

LOW   (We are responsive to our clients.)   HIGH

How do these responses match up with your previous surveys?

IT PROFESSIONAL ASSESSMENT

To further help us establish a baseline of understanding as to how our client base sees us right now, we should complete the following exercise. This exercise will also help us to establish how much or little work is facing us in meeting our service-oriented goals.

Honestly—how are we at:

☐ Timeliness of delivery of our services? _____

☐ Managing our cost of services, whether visible or as part of a charge-back system? _____

☐ Responding to our clients? _____

☐ Maintaining a response time that is consistent?

_____

☐ Meeting, greeting, and speaking to our clients in a manner consistent with the level of technology understanding our clients have? _____

If our assessment shows that our clients as a group see our service as merely "acceptable," we don't want to stop there! We want to deliver "exceptional" service! Ask yourself: Is this the service level that we as an organization want to deliver? Is it possible we could do better? Does what we are doing match or exceed what the competition is doing? Are there clients we could win back by doing a better job?

NOW WHAT?

We now know what we have to do. The challenge is to hold ourselves accountable and do it—without making excuses. A man by the name of Robert Townsend once said, "An important task of a manager is to reduce his people's excuses for failure." I agree—and would add only that each of us has the responsibility as an individual to reduce the number of excuses we accept from ourselves!

Our own ability to find excuses for anything that might possibly go wrong is one of the biggest obstacles to our success. We must constantly remind ourselves that the goal is never to find out who, other than us, is really responsible for a problem, and it certainly is not to find a way to make the customer somehow responsible for a problem. *This does not work.* Customers don't buy into our excuses. Instead, they lose faith in us. If they lose enough of that faith, they will simply leave us for the competition. We must resolve, first and foremost, then, to leave the "blame game" behind—forever. That means even something like, "We're sorry—Bob is not yet trained in this area. It's not his fault; he's doing his best"—is strictly off-limits.

No excuses! Whatever the problem may be, until we have fully investigated and know without a doubt the real answer to the problem, we must never put the blame on another member of IT. Of course, it should go without saying that we will never look for reasons to put the blame on the customer.

We are all in this together. We will work together as a professional IT team, we will work with the facts we have, and we will find a way to move forward … on behalf of the customer.

NINE COMMON MISUNDERSTANDINGS THAT CAN KEEP US FROM BUILDING AND IMPLEMENTING A SUCCESSFUL ONGOING SERVICE STRATEGY

Here are the nine most common misunderstandings that are circulated within IT about our service to our clients.

Misunderstanding #1: It is near impossible to try to figure out what clients really want and what their real needs truly are.

It's not impossible—it just requires a little effort and the willingness to speak the client's language instead of reverting automatically to ours. That effort to communicate without jargon is one of the elements that can—and, if we are steadfast, will—distinguish us from the competition. If it's impossible for us to speak English, avoid jargon, and ask good questions, then we deserve to lose our clients to the competition. This is the perfect place to get some practice with the listening model I shared with you in earlier pages.

Misunderstanding #2: *If they're not satisfied, it's not our fault. They should understand that we did our best with the resources available to us.*

Wrong. If they're not satisfied, it is our problem—and our responsibility. And P.S., in the vast majority of cases, we *haven't* done our best to resolve a problem. If a member of our immediate family called us with the same problem, would we shrug our shoulders after a single attempt and say, "Hey, we did our best"?

Misunderstanding #3: *Our clients tend to be unreasonable, and it is very hard, or even impossible, to satisfy them..*

If we take just a little time to build a better relationship with clients, if we make a genuine effort to better understand them as people with business pressures and needs, we will not only find it easier to support them—we will find it easier to understand them and work with them.

Misunderstanding #4: *Providing great service will only produce more work for me (or for our department), and no one seems to care if I am buried with work.*

Welcome to the new economy. You have three options: A, quit; B, watch your department get downsized or phased out; C, put your project requirements together, ask for a sit-down meeting with your superiors, and request more help, added resources, or more time. Here's my advice: Pick C every time, and don't be afraid to demonstrate the value you deliver—and show how your good work has brought on the "problem" of having too many people to serve!

Misunderstanding #5: *Change is not an option for us. We already provide great work..*

Change is a better option than a pink slip. I've seen organizations that believed this about themselves do a complete turnaround and become their companys' technical services providers of first choice. That having been said, it's certainly true that change will be easier for you to manage if your organization has a participatory management style in place (as opposed to a dictatorial management style). Sometimes, you must model the change you want to see take place in your organization.

Misunderstanding #6: *We don't have to overhaul the system—we can achieve the same service improvements with quick fixes.*

Want to bet?

Misunderstanding #7: *The cost of implementing this service philosophy is simply too high, and we hear there is no funding available.*

Only when you measure in the short term. Once you implement the new philosophy as expressed in this book, however, the payoffs on your investment—both measurable and intangible—will be immense and long-lasting.

Misunderstanding #8: *Top management is very busy and will eventually lose interest in this improvement of our service..*

Not when they start seeing the results.

Misunderstanding #9: *It would be so much simpler if we just left the client out of our decision making.*

Actually, this is not only a misunderstanding—it's also an attitude that destroys our reputation. If we didn't have any clients, we would no longer be in business. That may be "simple"—but it's not what we're after.

Disservice Attitudes

Now that we've exploded the most common service misunderstandings, let's take a look at the "disservice attitudes" that can also keep us from implementing great service.

Being DILLIGAD. As we now know, "Do I Look Like I Give a Damn?" never works well as a service provider.

Being removed sends a cold message and keeps clients at a distance.

Being "I know best." "Maybe if I show them how knowledgeable I am and talk down to them, they'll leave me alone." ("Forgive me. This may be complicated to you …")

Being rote. Applying the insincere automatic response: "Nice to see you, have a nice day, but I cannot help you."

Being a policy enforcer. "Sorry, the IT policy prevents me from doing that. Now, if it was up to me…"

Being "Find me." These folks avoid anything or anyone that may put any blame on them for anything.

Being "Not my fault." These folks always have some excuse for anything that may have gone wrong—and never see any possibility of correction.

Being on autopilot. This means giving mechanical responses and showing very little compassion for the client.

Being "I am busy." "If I don't make eye contact, maybe they will go away."

Being "Leave a message." "I cannot be reached on a personal level, so please leave a message on one or more of my many technical devices."

All of the above would prevent your organization from providing great service, but as you can see, these disservice attitudes can be eliminated. If we are really serious about eliminating them, we must:

* Not promise any more than we truly can deliver. In the old days, we used to say, "Oh, that feature will be in phase …" I hope we have learned that lesson by now.
* Assume an attitude of "We own the problem."
* Fully train our staff to understand all the components of service.
* Give each member of the staff full responsibility for all aspects of client service involvement, including attempts at problem resolution with the expectation of management support.
* Learn to back off or move forward by effectively reading the vital signs (visually and verbally) during each client interaction.
* Build measuring tools. You'll find them at various points throughout this book.

The Service Connectors!

There are many attributes that will guarantee a solid relationship with your clients. Here are a few of them:

When you are:

informed…responsive…professional…passionate
stimulated…interested…caring…conversant…

involved…concerned…excited…attentive… authentic…

Follow these and you will leave each client meeting well connected and respected each and every time… This I guarantee!

START RIGHT!

BEGIN EVERY CLIENT
RELATIONSHIP AT THE
DECISION-DRIVEN
LEVEL!

BUILDING YOUR CLIENT RELATIONSHIPS—ONE CLIENT AT A TIME

CHAPTER FOURTEEN

ONCE YOU KNOW HOW TO build "person-to-person" rapport with individual clients and prospects, you've got access to one of the very *easiest* ways to build, and keep, a competitive advantage. This advantage works on behalf of your employer, your IT team, and *yourself as an individual.* Fortunately, it's not all that hard to create and maintain good rapport with prospects and customers—which means that there's really no excuse for us not to do this. All the same, nine out of ten IT teams (using a *very conservative* estimate) fail to establish any kind of meaningful rapport.

How do we go about making sure that we're in the "right" ten percent on this issue? That's pretty simple: by being open to the idea that we should learn more about people. In my book *How to Market the Value of the I/S Department Internally*, published in 1992 by Amacom, I argue that if you ever want to be successful in marketing any product, you must first know all you possibly can about your buying audience. Specifically, you must know their buying habits, their culture, their likes, their dislikes, their concerns, their demographics, and anything else that connects to the emotional landscape where they use, or consider using, your products and services. The more you know about your audience, the better you can target your products or services for a proper fit. Of course, that principle applies to the ongoing job of marketing what our IT division does.

Ralph Waldo Emerson once wrote, "All persons are puzzles until at last we find in some words or act the key to the man, to the woman; straightway all their past words and actions lie in light before us."

To those wise words, which I think should be posted in every room of every IT department on earth, I can only add the following observation: The more you know about your clients, the better your service will be—because marketing is simply (as I define the meaning of marketing) **creating an awareness of value** in the eyes of your client ... one client at a time. And in that endeavor, I want to remind you that there really is no substitute for listening to your clients and finding out what is actually going on in their lives. Here's where we have to stretch ourselves a little. Our goal is not simply to listen to our clients to identify all the concerns about what we're doing (or not doing) well from an IT perspective, but also to listen to them *as people,* so we can come a little closer each time to finding the "key" that Emerson was talking about, the "key" that illuminates, not just their technical requirements, but a little bit more about who each one is as a person.

No matter whom you are talking to, that person will not be truly comfortable moving forward with you until and unless you create an awareness of value.

> In fact, I define marketing as **creating an awareness of value in the eyes of your client.**

The reality you face is this: You cannot expect to create an awareness of value—or an awareness of anything else—until you build up some rapport with that person. This is a universal law of human physics ... and it's something a lot of tech people (notably, our competition) are likely to skip. That's why we can't skip it!

Suppose you're having a conversation with one of your clients, who happens to mention that her first child just celebrated her second birthday. The *next* time you talk to this client, before you "get down to business," why not ask, "So, how is your two-year-old doing?" Or, if it's been a while between conversations, you might say, "The last time we met, you had a very young child. How old is she now?"

Or suppose your client mentions that his son is graduating from Yale in a few months. Make a note of it! At the next meeting, ask, "How are the graduation plans for your son coming along?" Continue with: "You must be very proud of him" or "What are his plans?"

This kind of exchange will help you connect on a human level and begin to build a stronger connection with this client. If you can do so, I advise you to keep a little notebook, data field, or separate database that will allow you to keep track, not only of service requirements and product issues, but also things like birthdays, recent family events, and important milestones in this person's life. I have included forms to help you keep track of this in this book. *The more authentic person-to-person connections we can make, the better the job we will do of creating an awareness of value.* The two really are connected—provided that you are *sincere* in everything you say to the client. If anything you say along these lines feels forced or unnatural, it will do you more harm than good, and you will lose, not gain, credibility with every word you say.

Occasionally—very occasionally—you will find yourself working with the kind of client I call the "no-nonsense type." This person sees little or no value in "chit-chat," and you'll be able to tell that right away from body language, eye contact, and tonality clues. Honor that person's communication style by moving right into the business purpose you have been asked to focus on. *You must not, however, assume that most of your clients operate this way, even though a few probably do.* Most of us enjoy some kind of personal connection when we are with others, whether we are in or out of our work atmosphere. Connecting with clients one-on-one is a skill that you must practice, and if you give it just a little bit of practice, you will find that it's one you can perfect. Once you do this successfully, you will probably find that you enjoy it and want to continue to do it.

You can start the process very simply. For example, while talking with your next client on the phone, simply try starting a dialogue like this one:

You: Hi, Sue, Paul from IT calling about project X. How are you today?

Sue: Oh, don't ask, I have just returned from a meeting on Project Triple Whammy, and I'm now responsible for getting it off the ground with too few people, not enough time, and no budget.

Bingo! You have built rapport. Wasn't that easy? Not only that, you now have information that you didn't have before about what this person is trying to accomplish ... and you may even have some clues about how IT could help.

Or perhaps you hear:

Sue: Wonderful! Couldn't be better!

You: That's why I always like talking to you—you have such a positive attitude.

Again—it's mission accomplished. There's rapport in the discussion. That's your responsibility: creating and supporting rapport. Start connecting with your clients at the human level! It works ninety-nine percent of the time. Don't let the missing one percent prevent you from taking advantage of the benefits of connecting to your client base one person at a time. Specifically:

1. During the beginning of every client interaction, take time to build rapport.
2. Get a check on their current perception of you, or IT—and adjust accordingly
3. Make every effort to deliver what they have come to expect of you or to fulfill the request that they've made.
4. Follow through with checks and balances to ensure client satisfaction after your discussion.

What are "checks and balances"? These are failsafe mechanisms—"doublechecks" that are built into our process that help us make absolutely sure that what we think is happening really is what's happening. For instance, after a customer has a buying or service experience with Toyota, someone from Toyota follows through with the customer to ask about the customer's level of satisfaction with the exchange.

"The Basics" for Effective One-On-One Interactions with Your Clients

1. Know your client's full name and spell it properly and learn to pronounce it properly. Use no nickname until given permission to do so. This is paramount!

2. Make sure that you know the exact department that he/she works at. Learn the person's location and work responsibilities. It is also advisable, if you can, to learn a little about the successes and concerns of that department.

3. Double-check his/her phone numbers and find out who may respond to the call. Who will be calling him/her directly? A co-worker? Office administrator? This applies to fax numbers, e-mail addresses, texting messages, etc.

4. Clearly understand his/her rank and titles. Always use his/her earned titles until your client informs you of his/her preference.

5. Be cognizant of your past history with this client: long term, short term, past project activities, and their results.

6. You may also consider your clients' educational levels, honors that they may have received, sports they may have played, their military service, if they are married or not, their children's names (if any), associations that they may belong to (such as a country club, the local Rotary), or any special community work that they may be doing.

7. What has IT's relationship been with this client? If there is a problem with his/her IT relationship, what is the client's side of this dispute?

8. Is this client authorized to go outside of IT for his/her computer needs? If yes, what impact would that have on your budget if this client chose to leave? What ripple effect would it have on your client base at large?

In this book, I've given you data-gathering forms you can use to gather information about the good people your IT team is interacting with. Use them!

BUILD RAPPORT!

CREATING A ONE-ON-ONE CONNECTION WITH THE CLIENT IS YOUR *SIMPLEST AND BEST COMPETITIVE ADVANTAGE* !

SUPPORTING THE TEAM

CHAPTER FIFTEEN

So—WHERE DO WE GO FROM here?

Let's start by reviewing the top action items that will support our team—the items that will keep us in a valued position with our clients if we execute them consistently, each and every time we have the opportunity.

FOUR CRITICAL ACTION ITEMS

1. At the beginning of every client interaction, take time to build rapport.
2. Get a check on the other person's perception of you, or of IT in general, and adjust accordingly.
3. Make every effort to deliver what the client has come to expect of you or the project or request that he/she made.
4. Follow through with checks and balances to ensure client satisfaction.

THE FORMULA

Now, let's go a bit further. To help us bring all this into perspective for the team, I have devised a formula that can help us see what we have to do in each and every situation to maintain a high level of service for everyone.

$$S +/- E (T) * MOT = RSL$$

This formula breaks down as follows.

Your service (S) as seen or understood by our clients. That service level can be perceived as …

Plus or minus expectations (+/- E), balanced by our clients' …

Tolerance (T). Some clients are more demanding than others, some are more forgiving than others, and some simply accept things as they are. Our job is to try to understand our client's tolerance level and adjust as required. At this point, our service level is further assessed by our client, based on the …

Moments of truth (MOT) that have occurred during the service delivery. Each MOT experience that falls below expectations reduces the value of the end service delivered; each MOT experience that exceeds expectations enhances the value of the service delivered. Combine all of these elements, and you will have established the …

Received service level. (RSL) This is the subjective level of service delivered as understood from your client's point of view only.

EIGHT WAYS TO MANAGE CLIENT EXPECTATIONS

Once we understand this formula and start utilizing it daily in our client interactions, we will achieve a very constant high level of client service satisfaction. This is another way of saying that managing client expectations is key to success. How do we do it?

1. Really get to know our clients.
2. Ask ourselves, "What would I expect if I were the client?"
3. Ask ourselves, "Can we manage that level on a consistent basis for this client?" If not, we must address this with our client right then and there—or we will set service limits that we cannot meet. The result will

be an unhappy and dissatisfied client. *We must always be honest and up-front as to what we can actually expect to deliver.*

4. Help our clients feel secure while working with all aspects of IT.

5. Don't be afraid to set the agenda for our discussion so we can be clear about exactly what can and can't be delivered.

6. Deliver more than they expected, whenever we can safely and ethically do so. Tell clients that this was only attainable because of special circumstances A or B or C ... so they will know that we may not accomplish this every single time.

7. Keep in close contact with our clients and keep them in the loop relative to their projects. Seek and respect their input.

8. Work hard at maintaining an open, trustful dialogue with our clients at all times ... and encourage them to complain—so we know exactly what the problem is!

BUILDING A STRONG AND LASTING SERVICE TEAM

Now that we have established what service really is and how to deliver it and gain its rewards, we should look at exactly what it takes to build a strong and lasting service team. Let's accept that our client survey has been returned and it clearly demonstrated that we need service improvement, and the sooner the better. Now that we have established the need, the value, and the cost justification for our organization to build our service team, what should we look for as far as characteristics and personalities that will bring us success right from the get-go? Well, many times, when we are confronted with this type of decision, we look around at our co-workers that we have established friendships with over time as nice and reliable people. And we tend to pick them.

Would you bet money on this approach working? I wouldn't.

Let's get a little more specific about the kind of team we will need to meet or exceed our client service goals that we have established and keep those goals at high levels for a sustained period of time. Below is a summary of some of the key questions we will want to address.

- What type of structure will we need here?
- What and how many positions will we need?
- What will the reporting structure look like?

- To whom will people report, and why that person?
- What level of authority will these positions have?
- What will be the pay and growth structure?
- Where will the funding come from?
- If we're launching a startup or a new division, should we have a fully-staffed, well-trained organization in place *before* our first day of business? Or should we start small and organize and add staff as required?

These are all critical questions, and we must address each of them directly with the top management of our company.

Let's assume that we have gotten top management to buy into the value of providing great service from IT.   With this in mind, let's look at the personality traits required to staff a successful, full-service-oriented IT organization. My list follows. This, I believe, is the list we must hire against, and train against.

We need people who …

… show an ability to be good **listeners.**
… show an ability to be **tolerant.**
… are **ethical.**
… are **visionary**—capable of seeing beyond the current problem.
… possess practical **intelligence** at multiple levels.
… are not afraid of taking **responsibility.**
… are **self-motivated.**
… are **respectful** of others.
… are **non-abrasive**—natural **people people.**
… are natural **team players.**
… have a great natural **attitude** (no DILLIGADs need apply).
… are able to accept **criticism** without taking it personally.
… are **neat dressers** and well-groomed.

Some people, after reviewing this list, have said, "Very nice, Ouellette. And I suppose we also want people who can jump over tall buildings in single bound, right?"

I would never try to tell you that all of these attributes must be in all of us 100% of the time. What I would say, though, is that these attributes are what is needed for us to function at a peak level as a service *organization.* The organization, we must remember, is the sum of all of our varied strengths and weakness.

Again—this is the list we must hire against, train against, and hold ourselves to *as an organization.* If we get this part down cold—and we can—we will be literally unstoppable. And we're probably a lot closer than you think. After some careful analysis about yourself and other members of your IT organization, you will conclude that Bob over there is known for his calmness, Peggy is very analytical and has a great way of presenting technical information to others outside of our department, and then we cannot forget Alfonse—he is the best team player that you have ever seen. And so on.

Once we know what service "personality traits" to emphasize as an organization, we will get the credit we deserve. And let me say, here and now, that we deserve a great deal more than we get. I personally have *never* seen a more committed group of people than IT professionals! When have you ever needed a swift kick in the posterior before starting any work-related efforts? Most IT professionals that I have met regularly work twelve-hour days. They keep checking in with the office for updates about their projects. They find themselves waking up at 2:30 am with a solution to their work problem. They try to write this solution down in the dark so they won't wake up their significant other. If they do wake up their significant other in the process, and they are asked, "Are you okay?" they say something like, "Everything's fine—I just went to the bathroom. Sorry to disturb you. Go back to sleep."

That's the level of commitment we already have. All we really have to do is make a few simple changes that will make it easier for us as an organization *to broadcast our commitment to our clients.*

Finding good service people is just as easy as looking a little closer at our current staff. Weed out the ones that simply would not be great service providers for some known reason, and with the majority that remains, build the best service team that has ever been assembled within your company's four walls. *Train* and *reinforce* the people who remain. The soft skills that support the principles I have shared with you in this book can be mastered and sustained with relative ease.

There are other skills to consider, as well. Good communications, good ethics, and good negotiating skills are all part of the service equation, and are all essential parts of the organizational "recipe" for getting, motivating, and keeping good people. We must constantly train and reinforce our team members in these skills. Even if we've "covered it all before"—we have an obligation to "cover it" again, and on an ongoing basis.

After just a few months of running our little IT business the way we've been discussing it in this book, you and I will think to ourselves, "Why on earth didn't we start this a whole lot earlier?"

On the following pages, you will find some more forms designed to help you ...

a) Develop satisfied clients.
b) Learn more about clients.
c) Support long-term relationships.

Look at the forms that follow—and use what makes sense for you.

DEVELOPING SATISFIED CLIENTS

- It's in everyone's nature to work first for things that have a reward attached to them. What could you use to reward your clients?

  How: _____

  _____

  _____

  _____

- Create a reward system in your department when superior service is delivered.

  How: _____

  _____

  _____

  _____

- See the value of creating a stimulating, positive work environment and employees who want to go to work in the morning.

  How: _____

  _____

  _____

  _____

LEARNING MORE ABOUT OUR CLIENTS

Tools Avalailable

Internal to our company:

- Websites, newsletters, brochures, departmental visits
- Other _____

_____

- Interviews, personally or within groups
- User groups
- Surveys, formal or informal
- Professional associations they may belong to

Questionnaires
- Reaction cards
- Pre- and post-service inquiries
- Current office communication about client situations
- Asking the clients about their level of satisfaction
- IT client surveys

Any other possibility?

_____
_____
_____

METHODS USED TO LEARN MORE ABOUT YOUR CLIENTS

Sample Organizational or Departmental Contact Information

Date: _____

Revised: _____

Organization's Name (or Dept.): _____

Key Players and Titles:
_____
_____
_____

Important Phone Numbers: _____
_____

Email:_____

Blogs:_____

History of IT and Its Service:_____
_____
_____
_____

Organization's Current Work Responsibilities: _____
_____
_____

The Department's Short-term Goals: _____
_____
_____

Current Concerns:
_____
_____
_____

Sample Methods Used to Learn More about Your Clients

Sample of Individual Client Information

Date: _____

Revised: _____

Name: _____ Title: _____

Address or Location: _____

Phone: _____ *Fax:* _____ *Email:* _____

History of IT Service with that Individual or Department:_____
_____
_____
_____

Organization's Individual Current Work Responsibilities: _____
_____
_____
_____

*Short-term Goals for this Client or Department:*_____
_____

CurrentConcerns:_____
_____

Marital Status: _____ Spouse's Name: _____

Children's Names: _____
_____
_____

Honors Received: _____

More Ways to Help with Client Dialogue

Sample Conversation Log

Time / Date of Conversation	What was discussed, requested, or promised	Dates of Follow-up Action

NOTE: This may be filled out while you are having a telephone conversation with your client or after an email or a visit.

Your Client Harmony Log

Client Name: _____

Improvements Made		Potential Negative MOTs	
Date	Description	Date	Description

EVALUATION

IT Liaison

One of our services is providing you, our valued clients, with a liaison to understand your needs and share IT strategies and directions. We want your feedback on your liaison and his/her service. Thank you for your time.

1. Does the IT liaison meet with you at least once a month? Yes_____ No_____

2. During that meeting, did he/she ask specific questions about your business needs?   Yes_____ No_____

3. How were you able to tell he/she has listened to your needs and worked with them?

_____

_____

4. Does the IT liaison keep you informed of IT changes? Yes_____ No_____

5. What can we do to make this relationship more valuable?

_____

_____

6. We are considering rotating IT liaisons every twelve months as a training opportunity. What is your reaction to this?

_____

_____

7. How can we make it easier for you to share information with the new IT liaison?

_____

_____

L. Paul Ouellette

8. Do you have any additional comments or suggestions?

_____

_____

9. Would you like a call back?   Yes_____ No_____

Name:_____ Telephone #:_____

EVALUATION OF IT HELP DESK EXPERIENCE

You called IT

You called IT on _____ and spoke with _____ about
_____.

We are interested in your opinion concerning the service you received
from the help desk.

1. Our commitment is to answer all calls in less than 15 seconds. Did
   we do that? Yes_____ No_____

2. Was the person to whom you spoke courteous, and was his/her
   manner professional?   Yes_____ No_____

3. While on the phone with IT, were the questions that you were
   asked …

   a. Not understandable or confusing? Yes_____ No_____

   b. Clear, to the point, and understandable?
       Yes_____ No_____

   c. Within your general/technical understanding?
       Yes_____ No_____

4. Did the resolution offered by the help desk work? Yes_____ No_____
   Comments:

   _____
   _____
   _____
   _____

Would you like a call back?   Yes_____ No_____

If yes, Name: _____ Extension: _____

EVALUATION OF IT SERVICE FROM THE IT CLIENT

We are interested in your opinion concerning the service you received in regard to the initial contact on _____ with _____for IT service._____

1. What led you to call the person you called?

    _____
    _____
    _____

2. Were you able to reach the person easily?   Yes_____ No_____

3. During your first conversation …

    a. Was your idea well received?   Yes_____ No_____

    b. Did you feel listened to?   Yes_____ No_____

    c. Did the questions you were asked make sense, and were they related to your initial request?   Yes_____ No_____

    d. How could the information gathering have been more effective?

    _____
    _____
    _____

    e. Were you given adequate information on the next steps? (e.g. magnitude, feedback, risks, your role, etc.) Yes_____ No_____

Comments:

_____
_____
_____
_____
_____
_____

SUGGESTED ACTION PLANNING FORM

Team	Individual

INDIVIDUAL COMMITMENT SHEET

Ask your IT staff to individually commit to their first step as service providers by completing the following:

Action Item	I plan to take the following action
I will meet individually with my top clients and ask about their major business concerns.	_____
Assist with my clients' business concerns.	_____
Congratulate my colleagues for delivering great service.	_____
Take some clients inside IT to meet my staff.	_____
Set up a lunch with a client to have a business free meal (simply lunch).	_____
I will circulate articles on great client service to my staff.	_____
Seek ways to improve my service levels.	_____
Plan to get on the calendar of my client's meeting.	_____
I will read business journals that my clients read.	_____

INDIVIDUAL COMMITMENT SHEET

Ask your IT staff to individually commit to their first step as a service provider by completing the following:

<u>**Action Item**</u>   <u>**I plan to take the following action**</u>

Within the next WEEK :

I will make customer service
my goal.   _____

I will call some clients and ask
how I could improve my service. _____

I will send some personal
thank-yous to clients for the
benefit of working with them.   _____

I will stay away from promises
that I cannot keep.   _____

I will start a daily ritual of
writing a plan based on my client
service for tomorrow.   _____

I will set time to visit my clients. _____

I will talk to five colleagues about
delivering great customer service. _____

I will further educate my clients
about technology.   _____

ASK YOURSELF!

WHAT IS THE CLIENT'S *RECEIVED SERVICE LEVEL?* THAT'S THE LEVEL OF SERVICE *HE OR SHE BELIEVES* WAS DELIVERED!

REAL-LIFE SERVICE
SUCCESS STORIES

CHAPTER SIXTEEN

I *HAVE MANY STORIES THAT* clearly show what providing great service can achieve, both for the practitioner and the recipient. I have selected three to share with you as we close this book. I chose these three events because, taken together, they identify the basic service requirements of any organization, regardless of its size, culture, or industry. The lessons these accounts share are applicable in every business setting, at any time. Every organization can learn from these examples—including yours.

"BUT I'M JUST A WAITRESS!"

I was on a three-day assignment with a major client in downtown Atlanta, Georgia. One of clients said, "Paul, we want to take you out for lunch today. We think you will really like what you'll see."

I said, "What? They show you the dessert first?"

The client said, "Just trust me on this one."

I said, "Okay."

Lunchtime rolled around, and he and a couple of his buddies took me across the street to the lunch counter at Macy's. The lunch counter, I learned, was closing down the very next day. I remember thinking to myself that I liked lunch counters, and I was saddened that they are fading away in this country. I expected to have a BLT and a soda as a final tribute to this lunch counter—I thought that would pretty much be it. Instead, I learned what service is all about.

The lunch counter itself took the form of three big, curved waves with one waitress working the inside of each wave. We arrived at the first wave with plenty of available seats, so we all sat down to order. As I looked for my menu, I happened to glance across the next wave, where I saw many empty seats. The third wave, however, left me awestruck. It was full of people, with a *long* line flowing against the wall with people waiting to take their seats.

I asked my clients, "What are all these people doing lined up against the wall? There are plenty of seats over here."

The answer came quickly: "They want to be served one final time by that waitress!"

Not only were the diners waiting, but most of them were carrying different types of gifts to present to this waitress: chocolates, flowers, that kind of thing.

"It has been this way all week now," my client continued. (This was Thursday.) "The governor himself is coming for his last lunch with her tomorrow."

I was totally blown away. I began to wonder: what is really behind this outpouring of affection for this particular waitress? I started to watch everyone's behavior at her station. I listened to her as she interacted with her patrons. And I asked my clients what made her so special.

Here's what I learned. She always made each meal an event for her patron. She always built rapport. She always made them feel as though she was truly blessed to have them as a customer. She remembered names. She'd ask questions like, "How's your daughter doing in college?" If anything went wrong with the meals (remember, we are talking about quick lunches here), she would always make the correction a *priority* and make sure everything was put right—without excuses or blame or judgment.

She went out of her way to be pleasant, to be sincere, and to create a connection with her clients. She did her job to the very best of her abilities, every time. When she bid a patron goodbye, she did so with full emotion—she meant it! She made even a quick chili meal a major event.

In short, she implemented every idea I have shared with you in this book.

A while later I conducted a service program for employees of this same client. By talking to them, I was able to find out where this waitress lived. I sought her out that evening, met her and her family, and invited her to my program the next day. As I was giving a speech on the benefits of providing great service, I had the honor of telling her story—and introducing her to the audience. I asked her for some advice as to how one can be a great service provider, and she answered, "Love whatever it is that you do, be nice to all people, and they will return the favor."

To that I say, *amen!*

If she can garner that much respect and admiration as a waitress, think about what we could accomplish in our organizations ... just by following her example.

THE CELL PHONE REPRESENTATIVE

A friend of mine got an unpleasant surprise after he opened up his monthly cell phone bill. After the cell phone company had helped him to select what seemed like the "right plan" for his usage level, he found that he had gone over his monthly allotment of minutes ... to the tune of nearly seven hundred dollars.

After spending more time than he really wanted to perusing the fine print on the bill, he called the cell phone company—and was lucky enough to get Kevin on the line. He explained his situation to Kevin. He also explained that, if he was supposed to expect monthly bills that were six hundred dollars higher than he'd anticipated, he wouldn't be able to keep using that carrier.

Here's the great part. Instead of hiding behind the policy and simply telling my friend that he really had used all the minutes that were listed on the bill, which is what a lot of telephone service representatives would have done, *Kevin apologized for the inconvenience and said he would look for a way to solve the problem.*

My friend nearly went into a state of shock when he heard that one.

While my friend was still on the line, Kevin found an unlimited-minutes plan that my friend could adopt for the same monthly price my friend was already paying. Then he said he was going to try to arrange a credit for the six hundred eighty dollars my friend had incurred in overage fees. Kevin explained that he had a limit of three hundred dollars that he was authorized to give in credits to resolve problems like these—but he also said that he was going to review the situation with his manager and make sure he was able to secure the additional three hundred eighty dollars that would resolve the overage fees entirely. If there was any problem, Kevin would make sure that someone from the company contacted my friend personally,

Shocked, my friend thanked Kevin for his help … and smiled. He had expected a major conflict over the bill, but because Kevin had done everything right—apologizing, accepting responsibility, resolving the problem to the best of his own ability and authority, and then escalating the issue to his manager in order to make sure it was fully resolved—my friend was now happy to be a customer of this cell phone provider. He ended up receiving a full credit for the overage minutes he'd used … and you can bet he told the story of his exchange with Kevin to lots of his friends!

THE KITCHEN CONTRACTOR

Many of us have dealt at one time or another with home remodeling contractors, and I'm sure many of us have stories to tell about those contractors—lots of them negative. This one, however, is positive.

My better half, Linda, and I decided to remodel not only our kitchen, but also three bathrooms—approximately 1500 square feet of space. After much research, we zeroed in on a company called The Granite House in Ft. Myers, Florida. The owner of the company and general contractor for this project is a man by the name of Virgilio Borges.

As you know if you've ever had to oversee a project like this, whenever anyone starts to tear something apart with the plan of rebuilding something, most anything can go wrong and probably will. Well, we hit a couple of bumps. Our general contractor simply kept repeating, "Don't worry—we'll take care of it." Both Linda and I were a bit skeptical, but we went on with the project and took Virgilio at his word.

Three days into the project, with most of the new kitchen floor completed, Linda noticed that the marble did not seem like the Travertine marble that we had ordered. She called Virgilio, who upon close inspection said, "Well … this *is* the tile that you chose in my showroom." Linda and I went to the showroom to see the tile in question and could not truly say, looking at it on the floor, that it was not the tile that we first saw. Maybe he was right. We knew, however, that it was not what we wanted now.

Virgilio suggested that we go look at a church that he had done the flooring for and for us to take a look. We did, and it was indeed a beautiful floor—but not the present tile that we had on our floor. He checked to see if we could find those same tiles for our floor, and of course, there were none around, and to order a new volume from Italy, where these tiles came from, was a costly and difficult proposition.

He suggested plan B: an upgraded Travatine tile that was still available domestically. He showed us samples. Linda and I agreed. He ordered the new tiles, ripped out the tile that we did not want, and proceeded to install the new tiles. He didn't charge us an extra penny and never said anything about the matter besides, "We want you to be happy."

He continued to greet us favorably throughout the project and continued to be very creative in resolving other problems that we encountered during the remainder of the project. (As it happened, there were unexpected and unavoidable problems.)

Virgilio put us first … just as that cell phone representative put my friend first … just as that waitress put her patrons first … and just as I hope you will put your clients first, as you implement the concepts I have shared with you in this book. If you do that, you will not just win the business … you will win hearts.

www.ingramcontent.com/pod-product-compliance
Lightning Source LLC
Chambersburg PA
CBHW051238050326
40689CB00007B/982